DANIELA ATENCIO

ROBOTIC TRANSLATIONS

DESIGN PROCESSES —LATIN AMERICA

COVERS:
ROBOTIC CIRCUMSTANCES: DESERT Z EXHIBITION. MATERIAL CLOSE-UP DURING EXHIBITION. 2023

INDEX

ROBOTIC CIRCUMSTANCES: DESERT X EXHIBITION. MATERIAL CLOSE-UP DURING EXHIBITION. 2023

INTRODUCTION

DANIELA ATENCIO

This book is a compilation of many learnings from the last 3 years. However, I consider it to be the result of many other aspects that have expanded from previous questions and some uncertainties given the circumstances in which this journey began. In fact, this is one of the main concepts addressed in the book.[1]

The texts are conceived as bridges between different strands. Research with creation in an exploratory, disruptive, and transversal pedagogy that promoted the use of the robotic arm to formulate work processes, workflows of creativity and encounters with technology. Learn from design processes with the apprehensions of the different practices as open and permeable spaces. Looking at reality as a truth, but also as an excuse to engage it in debates on the teaching of architecture and to test different working hypotheses. This book is not conventional in the sense of becoming a compendium of disciplinary knowledge. In fact, it is disciplinary in the way it opens new paths, showing its own routes from the point of view of rigor, ethics, and the aesthetic ponderation of various subjects. But also, from the personal and shared reflection on the search for new working practices in our region. It is, therefore, an anomalous and valuable book because of the context in which it is inserted [Latin-America], but also because of the medium through which it achieves it: the robotic arm.

It is important to establish that in this book will be described the first academic experiment that built educational spaces for the architectural project in the formation of architects in Latin America from the undergraduate level, moving away from the myth that the formation of researchers (and creators) in many of our disciplines, is developed in postgraduate or fifth-level studies. On the contrary, it has embraced the mission of approaching and preparing research and creation from an early age, linking collaborative spaces, building a collective of creatives and creators in many discussions about the present and the future of our discipline, and in the search for collaborative networks with transformative visions for our context through technology. This transformative or transdisciplinary vision has positioned this space as a pioneer in the region and has raised the interest not only of regional collectives but also from global practices.

As a brief background, this project began in 2020 but the COVID-19 pandemic delayed the experience of working with the robotic arm. At the end of 2021, with the reactivation of some academic spaces, the robotic arm became the definitive effort to promote new challenges and demonstrate that many aspects of our academic exercise had changed. In this case, the encounter with a new agent/actor began to leave a robotic footprint, leaving behind an overwhelming virtuality. Our academic

1 See *Robotic Circumstances* and *Peter Testa Interview* presented in this book.

preoccupation immediately turned into a celebration of collective enthusiasm. Activating the robotic arm and moving it, almost like an experience with *Mary Shelley's Modern Prometheus*[2], implied an immediate inflection.

The research-creation project that initiated these questions and actions was *Programmed Materialism*, which laid the groundwork for explorations in computational thinking. These early experiments, shown in the chapter of the same name, generated a series of visual results that were exhibited at the Venice Biennale in 2021.

This exploration of how to program materiality generated a concern for the transfer of the capabilities found in digital objects to new physical devices, which at the end became the genesis of translations design processes. In this scenario appeared the robotic arm and the construction of a series of procedures and design challenges related to its operation. From these premises, the research used the robotic arm, machines, and hybrid representation to find new approaches to the architectural project, from experimental production and its translations from various technological mechanisms.

The conceptual apparatus designed to both operationalize these issues and give agency to the robot in the design process has been described in a workflow empowering translations as the link between five specific states or conceptual frameworks that determine a path, a variety of processes, but also a connection to other issues related to representation and advanced digital fabrication.

The exploration of representation unites these processes in a body of work that uses technology as a creative, open, reflexive, inclusive and plural agent. Creativity is the mean of thematic cohesions and the enunciation that promoted collective intelligences that were constructed in formal spaces such as the classroom and the laboratory, but that expanded into diverse, speculative, and often academically informal places. In other words, innovation in pedagogies, or pedagogies of innovation, are part of the process and spirit of feedback in this work between students and teachers. It has a hybrid, inventive and speculative character from a pedagogical and research-creative point of view through technology.

The aim was to investigate, identify and propose new approaches to architectural artifacts from different disruptive patterns, promoting a debate on the future of our discipline, always from the practice of research through design and learning by doing. There is also a description of hyper-artifacts as a definition of the research itself, based on what has been established by various authors, in order to accumulate and provide new capacities to the result of the design exercise in architecture.

The book is structured in three continents[3]. The starting point, which is the preamble of conceptual support for the experiments developed. In this space, aspects of the conceptual base are developed, but it is also positioned in the very space where these practices were determined (Latin America), giving support to some of the concerns developed empirically in the design processes.

2 Shelley, M. W. (1818) *"Frankenstein; or, The modern Prometheus"*. London: Printed for Lackington, Hughes, Harding, Mavor, & Jones.

3 Continents in the sense of encompassing a series of contents that have affinities conceptually or by processes. Continents are not intended to be understood as hermetic packages.

Gabriel Esquivel's prologue serves as a theoretical framework that helps to delimit some of the questions developed in the book, but above all, it functions as a gateway to diverse meanings around the robot, technology, and the relevant debate as an actor within the contemporary scenario, especially in the region. It is here (and from here) that it's going to be establish the definitions of translations, especially of robotics, hyper-artefacts, and some aspects of representation. Claudio Rossi accompanied me in various parts of this exercise, but also to expand the vision of the context in which the research was deployed. The selection and description of the case studies, as well as the relationship with aspects of the region from disciplinary positions, affirmed in the text Latin-America in Worm-Eye View, a series of points that are reflected in the work presented in the following continents.

The next continent frames the workflow and its different translations. In this space, the different operations, and explanations, sometimes conceptual, sometimes a little more technical or operational, have no other aim than to promote different observations on what has been learned in these years while explaining some personal experiences in the pedagogical work around the robot. It is important to emphasize that these design processes and workflows accumulate an important academic experience in the teaching of undergraduate architectural design.

The following continent is the result, not as "outcomes" but as evidence of different experiences that function as final reflections learned and lessons for the future. This, accompanied by further conversations about the performance with Peter Testa, frames the final effort as the book's closure.

Finally, a series of simultaneous images between the first design processes developed at the beginning of this journey in *Programmed Materialism* and the results of the robotic translations workflow and the last exercise executed in the Robotic circumstances exhibition. They are neither conclusive, nor reflective (in the sense of reflecting comparisons), they are understood as the two borderlines that could be crossed by the bridges built over the years.

This book was designed to be understood from to discourses: The first, the interconnection between different texts and authors disseminated along the written pages. Secondly, by its graphic visual content that can be also bridged (or not) with the written debate. Feel free to navigate through this book using these two parallels but connected discourses. It possesses a permeable structure that can be revisited in order or in disorder, to be understood as a conceptually visual and visually conceptual apparatus; it is an artefact, or a paper hyper-artefact made of pages filled with many lessons of its own, expectant of learnings for others. Another last bridge between what have been done and what can be do, especially in this region.

From a Translated Robotic Worm-Eye View position.

PROLOGUE: TOWARDS UNDERSTANDING ROBOTIC ONTOLOGY

GABRIEL ESQUIVEL

As a society, we are propelled to explore the relationships between instruments, humans, minds, and machines, as well as discourse and its praxis. We are constantly discovering that we are no longer modern precisely because the network of factors that make any act of knowledge possible is extremely more complex than the theorists of modernity imagined and because what it gives rise to is much less elementary than the truth. Furthermore, thus, from constructivism, we passed from pluralism to a sophisticated ontological pluralism.

A mechanical object is not a truth about the world, but human consciousness tends only to reproduce the psychology of the imposition of technological totalitarianism, even if we complain about it. The great myth of industrial society was that every human problem could be solved with an object. Today, within a post-industrial and post-digital society, things have changed. We tend to conceive the idea of automation with cultural aspects directly from science fiction. We should be greatly mistaken when we view science fiction as an escape from everyday reality; on the contrary, it is an extrapolation from the irrational tendencies of that reality through the free exercise of narrative invention.[1]

Traditionally, according to *Baudrillard*, human well-being continues to be reduced to the convergence between technology, production, and consumption[2]. The consequence of this productive order is that it supposes the existence of a moral handicap of the human being concerning technology. Thus, technology becomes the moral authority that dominates our activities, and human conscience is reduced to compliance with the operational systematics of technology. The world is produced, dominated, watched, and controlled, and human consciousness advances until convergence towards the same order. Within the post-industrial and post-digital societies, ideology must break with the essence that the world is produced and dominated, subjected to a total inventory that monitors and controls each process. Unfortunately, the widespread cultural conception regarding automation and artificial intelligence is still unclear. We could continue with the same myths and fears based on science fiction, preserving the structure of a consumer society.[3] Today's world hopes to reverse the consumption paradigm by production, where we can locate our relationship with robotics, automation, and artificial intelligence.

Science fiction has practically nothing to do with the real future of technological development, for which it accounts in the future perfect tense, so to speak, drawing for nourishment on sublime archaisms and a repertory of acquired forms and functions. It contains little in the way of structural invention, but it is an inexhaustible mine of imaginary solutions to stereotyped needs and functional requirements of an often marginal or mindboggling variety; unfortunately, this way of thinking is still prevalent in architecture.

Technological production is not objective since it is supervised by a social system of practices that direct this production. Technically, there is only meaning; the contradiction is found in the practical ideology of the system. The strength of the robot world is the human being's dominance over it. The itinerary laid down for the robots led inexorably to the complete automation of their existing functions and has far less to do with humanity's future technology than with its present cultural motivations.

Fundamentally, the robot is simply the mythological end-product of a naïve phase of the imagination, a phase which implies the projection of a continual and visible functionality. The substitution must be visible: if it is to exert its fascination without creating insecurity, the robot must unequivocally reveal its nature as a mechanical prosthesis (its body is metallic, and its gestures are discrete, jerky, and unhuman).[4] Automation is the myth of the modern object. The value of a particular object is its degree of automatism. Humans project their consciousness to objects; these are the ones that have life. It is "*the forced indifference, of the passive*

1 See Baudrillard, J. (2020) "*The System of Objects*". Translated by Benedict, J. London: Verso

2 *ibidem*

3 *ibidem*

4 *ibidem*

spectacle of their power."[5] Objects become independent of the gesture of human action until the manufactured world replaces the natural world. The environment becomes more technical, which produces a fear of the atrophy of human behavior.

This technical environment presupposes the submission of the human to the technical, to the absolute point of the mechanized society and the machine world. This way of conceiving the relationship between robots and humans is a total myth. It presupposes that the human being becomes dysfunctional as the object becomes more functional. The struggle we face in the relationship between humans and automation is critical in disciplines like architecture, where we tend to speculate even before we fully know or understand what any specific technology is or does. We tend to look at it through a cultural filter that is sometimes profoundly rooted in modern biased myths. Our technological civilization has a rule that techniques and objects suffer the same servitudes as human beings, and the process of material organization, hence of objective technical progress, is subject to the same blocks, deviations, and regressions as the concrete process of the socialization of human relationships, hence of objective social progress.[6] Robotics development has grown exponentially, so much so that we coexist daily with some types of robots. Robotics has dramatically advanced in designing support tools in various domains. However, for more complex problems and solutions, it must be enhanced with other branches of knowledge, such as artificial intelligence. Thinking about how robots enter everyday life, the need arises to find a support tool in this area; this is where the need for robotic governance is found.

By looking at interactions between humans and robots in a cross-cultural workplace and communication framework, we could examine robots as coworkers from different cultures. Is there a way to study robots and the phenomena we are interested in without attributing culture to the robots? Robot culture would somehow involve shifting the conversation's focus from the humanistic culture of people who act with robots to the culture of an equally valid robot ontology. Robotic ontology could be an uphill battle for humanists, something like what happens when the culture of robotics and manufacturing is transported to the academic environment in Latin America. These people are not against the idea but do not see the added value in attributing nonhumans to a culture. There are possibly better, uncontroversial forms of analysis in distributed cognition in a heterogeneous network of humans and robots.

Robot-Actor Theory

Understandably, there is a strong visceral reaction against the attribution of culture to the nonhuman sphere. It is simple to understand that we continue to operate within an anthropocentric culture. This non-human culture is the same fight we have to articulate with the opponents of digital culture that is somehow considered foreignizing. We find ourselves needing to use Bruno Latour's terms with our colleagues as 'agency.' There is a deep rooted assumption that human beings have certain innate, unique qualities that distinguish us from mere matter and other animals. When someone comes and makes a very nuanced point about how objects have agency, the most immediate and natural response is, first of all, anthropomorphism, which is easy to refute.[7]

Latour's ontology is flat, with all entities having the same capabilities. They have the same capabilities *a priori* but are different, especially after the socio-technical relationships arise and an operation begins. This means that contrary to common interpretations, objects do not have intentionality or consciousness because —and this is the critical point— neither do humans. The concept of an intentional object is central to the account of existence and nonexistence. Moreover, merely intentional objects, which exist only in being represented, pro-

5 Baudrillard, J. (2020) "*The System of Objects*". Translated by Benedict, J. London: Verso

6 See Baudrillard, *Op. Cit.*

7 See Latour, B. (2005) "*Reassembling the Social: An Introduction to Actor-Network-Theory*". New York: Oxford University Press.

vide the clearest paradigm of things that are but are not in themselves.[8] In another interpretation, perhaps humans have intentionality or consciousness, but it makes no difference one way or the other. A good actor-network theorist can take some existing systems and start exploring explanations based exclusively on human qualities and give an alternative explanation based instead on materials, technologies, infrastructure, documentation, and other modes of outsourced practices. It is not about creating a fictional narrative but thinking about the future of artificial intelligence and machine learning and how this interaction between human ontology and robotic ontology will occur.

Unlike traditional humanism, in which only human beings were recognized with the power of agency, in post-humanist theory, the capacity to convey agency is recognized in all living or non-living beings. It is not vain to say that a work of art or any other object captivates us. The verb captivate is the best indication that we are talking about agencies- the effect or power that something exerts over us. It attracts us like the force of gravity attracts bodies. In terms of agency, no absolute distinctions can be made between humans and nonhumans. That is why we also speak of distributive agency: The possibility of exercising agency should not be limited to its traditional bearers – human people – but should be understood as a property of all things. There is no single subject in the cause of an effect, but many vitalities always come into play.[9]

Latour makes his theory of the actor-network—in which he proposes symmetrical responsibilities between human and technical actants—from the asymmetries that can arise, depending on the context in which the human actants use the technicians. What is the place of human agency in the society network? We maintain that these conceptions correspond to the agency-structure axis of social theory or sociology. To begin, we need to investigate the similarities and differences between these approaches and the explanatory scope of the network concept of each one for social theory. When Latour attributes agency to objects, he manages to do so by keeping the agency of humans and the agency of nonhumans symmetrical. Against the standard objection that he attributes too many human characteristics to nonhumans, what is happening is that he is doing so much by eliminating many of the "uniquely human" qualities of human agents. This elimination is perhaps why *Latour* never goes inside anyone's head, and he rarely tries to give a psychological or cognitive explanation in the networks of actors he studies.

Ed Hutchins examines a set of phenomena that have fallen into the cracks between the established disciplines of psychology and anthropology, bringing to light a new set of relationships between culture and cognition[10]. The standard view is that culture affects the cognition of individuals. The author argues, instead, that cultural activity systems have cognitive properties that are different from the cognitive properties of the individuals who participate in them.[11]

Hutchins explores the computational and cognitive properties of larger systems than individuals. These properties could be interpreted as a robot having to consider its "agency within a cognitive but somehow cultural computational exchange.[12] It then moves on to analyzing learning or change in the organization of cognitive systems at various scales. His conclusion illustrates the price of ignoring the cultural nature of cognition, pointing to how contemporary cognitive science can be transformed through new meanings and interpretations. It is producing the effect of a heterogeneous network of human and nonhuman beings.

8 See Merrihew Adams, R. (2022) "*What Is, and What Is In Itself. A Systematic Ontology*" New York: Oxford University Press.

9 See Bennett, J. (2010). "*Vibrant Matter. A Political Ecology of Things*". Durham: Duke University Press.

10 Hutchins, E. (1996) "*Cognition in the Wild*". Massachusset: The MIT Press.

11 *Ibidem*

12 *Ibidem*

Robots: Legal Entity, Taxation, and Governance

Currently, science and technology experts assure that in the Second Digital Age, there is fear in Architecture and other disciplines that it will replace many jobs in the short and medium term. We may think that it is only a political and economic issue. However, when analyzing from a philosophical point of view, the society we want, how we want to live and work, how to interact with robots, and what role they will play in society. The idea of providing robots with a legal personality - since they will be more autonomous, more intelligent, and able to learn from their environment - requires reviewing who will bear the civil responsibility for their actions. This issue of robots becoming legal entities has sparked a broad and prolonged debate about science and technology and what they imply for humanity in the future, and this topic has not been addressed in depth within the discussions of architects.

Many analysts attribute part of the accelerated robotization process to the artificial effect of tax systems, which distorts the decision to hire robots at the expense of human workers. Taxation, such as payroll taxes and mandatory contributions to employers' social security, falls heavily on the hiring of the labor factor and makes the human payroll of firms more expensive. Although there is no such thing in the tax codes as a chapter on the tax regime for robots – among other reasons because it is a complicated subject to define from a tax point of view – robots are a form of capital.[13]

It is essential to explore the concept of robot standing as a helpful idea that can assist in the anticipatory governance of robots. However, simultaneously, the concept necessarily involves speculative thinking, as it anticipates a future that has yet to arrive fully. We must situate robotics in the context of anticipatory technology governance by emphasizing that robots are currently becoming objects of governance. The robot standing concept posits that artificial agents could have claims to novel forms of moral and legal status.[14]

Thus, it is closely related to discussions on new understandings of technological artifacts and related changes in the conceptualization of agency. This essay/introduction does not provide new conceptualizations beyond what *Daniela Atencio* and *Claudio Rossi* are already working on, or ideas related to the discussion on robot standing itself; instead, it reflects on the usefulness of having such a discussion. The speculative content of the robot standing concept is argued to be instrumental in constituting robots as objects of governance.[15]

Perhaps the main objective of this introduction is to continue the conversation about the philosophical implications of the development and convergence of humans, AI, and robotics today. In the background, there is an outline of the debate between humanists reluctant to technical advance and supporters of these new technologies. Although to understand the current technical dimension of the human being, it is also necessary to rescue the thinking that germinated at the end of the 19th century and continued throughout much of the 20th century about the impact of science and technology on people and social reality. This impact may be less important within the academic context of the most important universities in the United States, Asia, and Europe. However, the Latin American case has been different; there is a cultural identity implication in terms of a social and economic position and, within the medium of architecture, a fear of the unknown that comes with dismantling formal, material, and structural traditions of making and teaching architecture. All these issues make Daniela's work/research more important; she should be considered a true double pioneer as a technology definer and as a woman in a complicated context.

13 See Porter, E. "*Don't Fight The Robots. Tax Them*", New York Times; February 23, 2019, https://www.nytimes.com/2019/02/23/sunday-review/tax-artificial-intelligence.html

14 See Coeckelbergh, M. (2010) "*Robot Rights? Towards A Social-Relational Justification of Moral Consideration*". Ethics Inf Technol 12.

15 See De Pagter, J. (2021) "*Frontiers. Speculating About Robot Moral Standing: On the Constitution of Social Robots as Objects of Governance*". December 2, 2021. www.frontiersin.org - Institute for Management Science, TU Wien, Vienna, Austria.

Conversely, the generational distance between those who promote it and those who accept it is diluted between being stable or disruptive. In general, academic programs in the region need a policy to include systematized emerging technologies, which makes the incorporation process slow. The implementation in Europe and the US occurs in spaces where design problems are encouraged over the instrumentalization of a process. The implementation of technologies still exists in the region as an extension or continuation of the practice of drawing, and, as it is not integrated into the design, education is oriented towards technological consumption but not towards development or discovery.[16]

Conclusion

Atencio's work occupies interesting interstitial borders between humans and robots. However, those borders are not fixed, but constantly in flux. At a certain point, the attempts to differentiate a robot from a non-robot become arbitrary because there is a human tendency to delineate and classify boundaries between humans and other ontologies. The constant adoption of increasingly complex forms of architectural reality in all facets of contemporaneity requires a rethinking of human, digital, and post-digital collaboration within the design process and its method of representation.

In their article *Programmed Ruination*[17]*. Hyper Artifacts*, *Daniela Atencio* and *Claudio Rossi* describe their research around the technological approach and its relationship with the discipline of architecture. The design combination of specific technological tools (software and hardware), focusing particularly on the interaction with a 6-axis robot arm, deepens the exploration of design strategies that can amplify the pedagogies guiding undergraduate students in Latin America. Their idea of working in Latin America in the Post-Digital world explains the importance of abandoning early digital biases about economy, identity, and nationalism and embracing new forms of robotic governance. The ubiquitous nature of the digital world guides them to pioneer different forms of operation which requires a redefinition of robotic ontology that includes even economic benefits (taxation) which was never addressed before as such within Latin American architecture. This form of pedagogy, research, and creation process allows for strengthening and questioning disciplinary issues through new speculative and narrative techniques or conceptualizations of the architectural object.

That is why it is crucial to consider the relationship between humans and nonhumans in the case of robotics as an aspect fundamental to society's culture in the future. Despite the risks, consider that the solution is not found in the rejection of technology. We should demand a reflection on the emerging technologies of AI and robotics, incorporating an object into another object in the world. In short, it is an object that cannot be reduced to mere scientific objectivity. Currently, numerous academic research centers in architecture including Atencio's work are developing both practically and speculatively the technology to improve the education not only of the architects who are part of these research groups but of the society itself, both in theoretical and practical terms.

Why deny the new generations advances in science and technology? Here, philosophy is a fundamental pillar in critical thinking and consciousness of the past, present, and future concerning science and developing technologies. Therefore, the question is raised whether the emerging technologies of robotics and AI represent an improvement in life on the planet or the problems that come with this. The philosophy of technology has already made considerable efforts toward increasing involvement in developing policy ideas. The development of a concept like robot standing agencies philosophically, legally, economically, and in matters of governance should be understood as an effort to develop speculative but rigorous concepts regarding how we should view these concepts when it comes to architecture.

16 Herrera, P., Juárez, B. (2012) "*Perspectives in Latin America's Fab Labs*". Universidad Peruana de Ciencias Aplicadas, Perú. Espacios Digitales. Fab Lab Lima.

17 Atencio, D., Rossi, C. (2024) "*Programmed Ruination. Hyper_Artifacts*". Cit. In: Edited by Esquivel, E. (2024) "*Design Technology and Digital Production. An Architecture Anthology*". London: Routledge 2024.

ROBOTIC CIRCUMSTANCES: DESERT Z EXHIBITION. MATERIAL CLOSE-UP DURING EXHIBITION. 2023

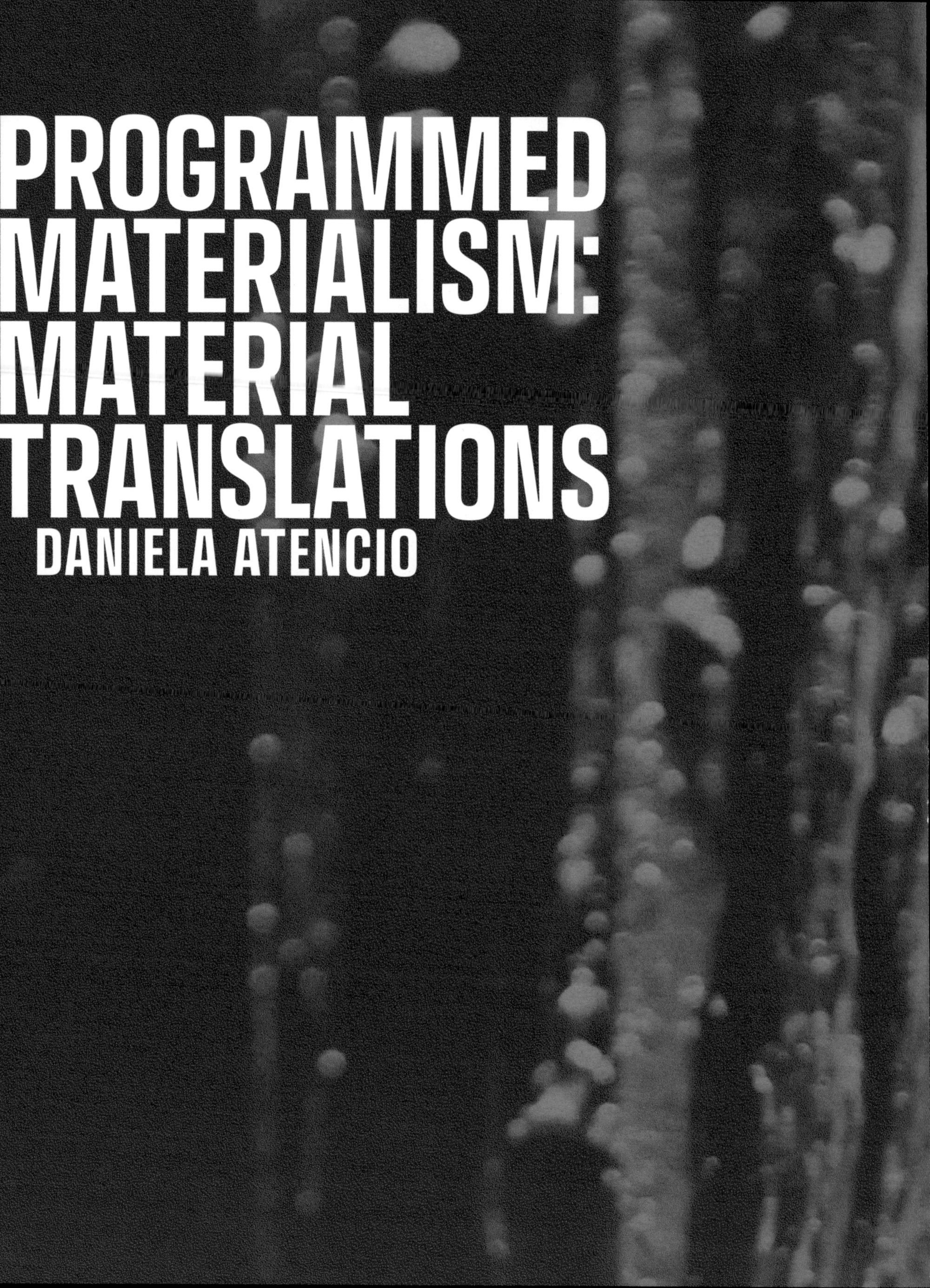

PROGRAMMED MATERIALISM: MATERIAL TRANSLATIONS

DANIELA ATENCIO

The processes named *material translations* started as part of the research *Programmed Materialism*[1] exhibited at the the *City-X Pavilion in Italy, Venice Biennale* 2021. The experiments of that exhibition lead to a reflection on how to involve computational thinking in the architect's undergrad and grad formation. The first *translations*[2] began to appear as soon as these exercises were formulated as processes (or procedures) of addition or subtraction, for example, recalling the compositional modes of Le Corbusier[3], among many other architects[4]. However, it was closer to understand the composition and decomposition of objects' thicknesses based on protocols, as some treatises on architecture of the eighteenth and nineteenth centuries. *Jean-Louis Durand's* treatise *Précis of the Lectures on Architecture*[5], which ordered his ideas from matter arrangements, to then establish design positions as assemblies, ensembles, applications and unions of elements and parts was pretty much the objective. In other words, the processes presented here do not lead only to a discussion of disciplinary exercises, but to generate valid argumentative bridges for the creative exercise of the contemporary architectural narratives and its possible forms of execution. After all, this process was the starting point for many experiments in the field of architectural education, its pedagogy and the possible methodological bridges based on technology deployed in this book.

This chapter focuses then on describing different processes of translation between images and material artifacts, as well as time-based procedures of manipulation and representation to inform matter. It comprises a series of methodologies that, beyond focusing on exploring novel forms and their readings, critically examined how digital tools can transcend visualization and advance the architectural discourse in the current second digital era through materially informed design processes. The investigation was then based on speculation processes of computational design as well as a critical-conceptual framework through instructed processing and parameters. It investigated the

1 Atencio, D. (2021) "*Programmed Materialism. Fictional Translations*". 17th Venice Architecture Biennale Italian Pavilion Sezione del Padiglione Italia—CITYX, Venice.

2 See the definition of *Translations*

3 Le Corbusier (1929) "*Tracés Régulateurs*" cit. in. Quetglas, J. (2009) "*Les Heures Claires. Proyecto y Arquitectura en la Ville Savoye de Le Corbusier y Pierre Jeannerete*". Barcelona: Asimétricas

4 Eisenman, P. (1998). "*Diagram: An Original Scene of Writing*". ANY: Architecture New York, 23, 27–29.

5 Durand, J.N.L. (1802-5) "*Précis des Leçons D'architecture Données à l'École Royale Polytechnique*". Paris: The Author.

ways in which meaning, intentions, and knowledge are built by solving and materializing complexities. Multiple prototypes and distinctive non-conventional objects were generated, manipulated, and translated into a graphic-objectual discourse as a computational argument through four connected material experimentations: *new materialism, material superposition, unlocking materiality, self-materiality actualization,* and a series of fictional scenarios (*Contextualizing De-contextualized Objects*) as visual conclusions.

1. New Materialism_Code

The core of this exploration was to develop a new form of representation of matter, seeking and deepening into the correlation and non-correlation between the real and the fake, the physical and the digital, the past and the future, the permanent and the temporary. The starting point was the materialism approach of *Bruno Latour*[6] combining the idealist definition of matter with its different agencies as a mixture of ontologies in a non-correlation philosophy, which can only be achieved by generating a new way of thinking and breaking certain parameters to which humans are familiar[7]. This procedure started with a reversed notational logic: vectorial extractions from images using code instead of plain images as the input. Therefore, a series of multi-iterations combining image resolution and information transformed images into a new digital matter, or, using *Latour's* definition, a *new materialism*[8].

2. Material Superposition_Add

Materialism as part of a non-correlation design, makes seem geometry and matter as a non-correspondence process. According to *Laruelle*[9], the digital representation and the 'object' are both materialism in a different way of thinking, working as an exchange and non-exchange conceptual superposition between these two common separated entities. This phenomenon can be shown in the works *Marcel Duchamp*[10], where objects often have a multiplicity of meanings, emphasizing the mutable nature of identity and, perhaps, demonstrating the interconnectedness of identity and the fluid or malleable nature of meaning, overlapping frames and multiple perspectives. Adding, superposing, or merging layers of meaning or a new composed identity of accumulative parts, was the speculative process of this material translation.

3. Unlocking the Materiality_Remove

Bruno Latour brings to the discussion the idea of unlocking the matter through the categorization and speculation of it, to let explore those qualities that are not accessible or visible but at the same time, arousing the intrigue to know the ways and qualities that may emerge from such capacities.[11] Removing processes through geometrical subdivisions and subtractions unlocked those invisible capacities, almost as a rediscovered digital archeology of the images and their translations into objects.

Unlocking materiality through a process of removal, subtraction, or fading of matter thicknesses made this experiment closer to a digital archeology than to material assembly processes.

6 Latour, B. (2007) "*Could We Have our Materialism Back Please*?", In Thick Things edited by Ken Alder. 138-142

7 See Laurelle, F. (1996) "*Principles of Non-Philosophy*", London: Bloomsbury Academic

8 Latour, B., *Op. Cit.*

9 Laurelle, F., *Op. Cit.*

10 This observation is made from the work of *Marcel Duchamp's* (1887-1968) 'readymade' objects such as the Bycicle Wheel (1913) and the best known of his works: Fountain (1917), submitted to an exhibition of the Society of Independent Artists in New York. *Duchamp*, one of the creators of the *Dada* movement, explained the theory of his 'readymade' objects in an anonymous editorial published in the May 1917 issue of the Avant-Garde magazine *The Blind Man*, saying: "(...) *he took an ordinary article of life, and placed it so that its useful significance disappeared under the new title and point of view - created a new thought for that object*". It is from this position and change in the point of view to produce a new definition of objects that its possible to rethink many of the disciplinary conventions.

11 Latour, B. *Op. Cit.*

In this sense, what was pursued was not only to 'undress' the materiality but also to rediscover new hidden capacities, arrangements, and protocols that allow us to give new meaning to the elements that compose it.

4. Self-Materiality Actualization_Physics

The constant rethinking of the representation[12] in our field of knowledge opens the question of the mechanisms used to explore the argumentative value of it, not only as a visual or digital phenomenon, but also as an exploration (and struggling) process itself. In the words of *Negarestani*: "(...) *to be human is to struggle. The aim of this struggle is to respond to the demands of constructing and revising human through the space of reasons"*[13]. It is characterized as developing a certain conduct or error-tolerant deportment according to the functional autonomy of reason; an intervening attitude whose aim is to unlock new abilities. In other words, it was to open up new frontiers of action and understanding through various modes of construction and practices (e.g.: social, technological, among others), but in this case, as new material capacities and frontiers, through a series of dynamic experimentations and actualizations with behaviors such as gravity, wind, resistance, collisions, orientations, and /or influences.

5. Fictional Translations: Contextualizing Decontextualized Objects

The fictional scenarios were the digital ('*fic-gital*') representation and materialization of all the processes. Perhaps one of the most exciting anticipated conclusions in this speculative body of work was the action of contextualizing decontextualized objects to open the debate on disciplinary issues usually associated with materiality in architecture: scale, orientation, fabricability, inhabitability, and trying to find adaptations and *translations* in a transdisciplinary, permeable debate necessary to reorient reflections in our field of knowledge. In essence, constructing fictional translations to achieve arguments through the exploration and experimentation of materiality in a 'feasible' scenario.

This last translation was a crucial exercise in understanding the previous explorations. Situating, implanting, locating, and providing architectural capacities promoted a series of reflections that were presented as visual conclusions. In this way, some of the questions developed during the process of translations between images and digital objects were revisited, establishing an architectural lexicon nourished by technological procedures. But it also opened new questions about the material capacities of these explorations. These capacities are pushed to the limit from a conceptual point of view, but also as initial questions for new protocols such as those proposed in the subsequent chapters of this book. The robotic arm will be one of the working agents that will help to initiate a new 'adventure' and continue others.

12 See *Off/On* chapter of this book, as weel as the Latin-America in Worm-Eye View where there are some positions about representation in architecture, specially in the Latin-American context.

13 To expand on this theoretical debate, see also *Reza Negarestani* (2014) "*The Labor of the Inhuman, Part II: The Inhuman*" in Journal E-Flux, 53.

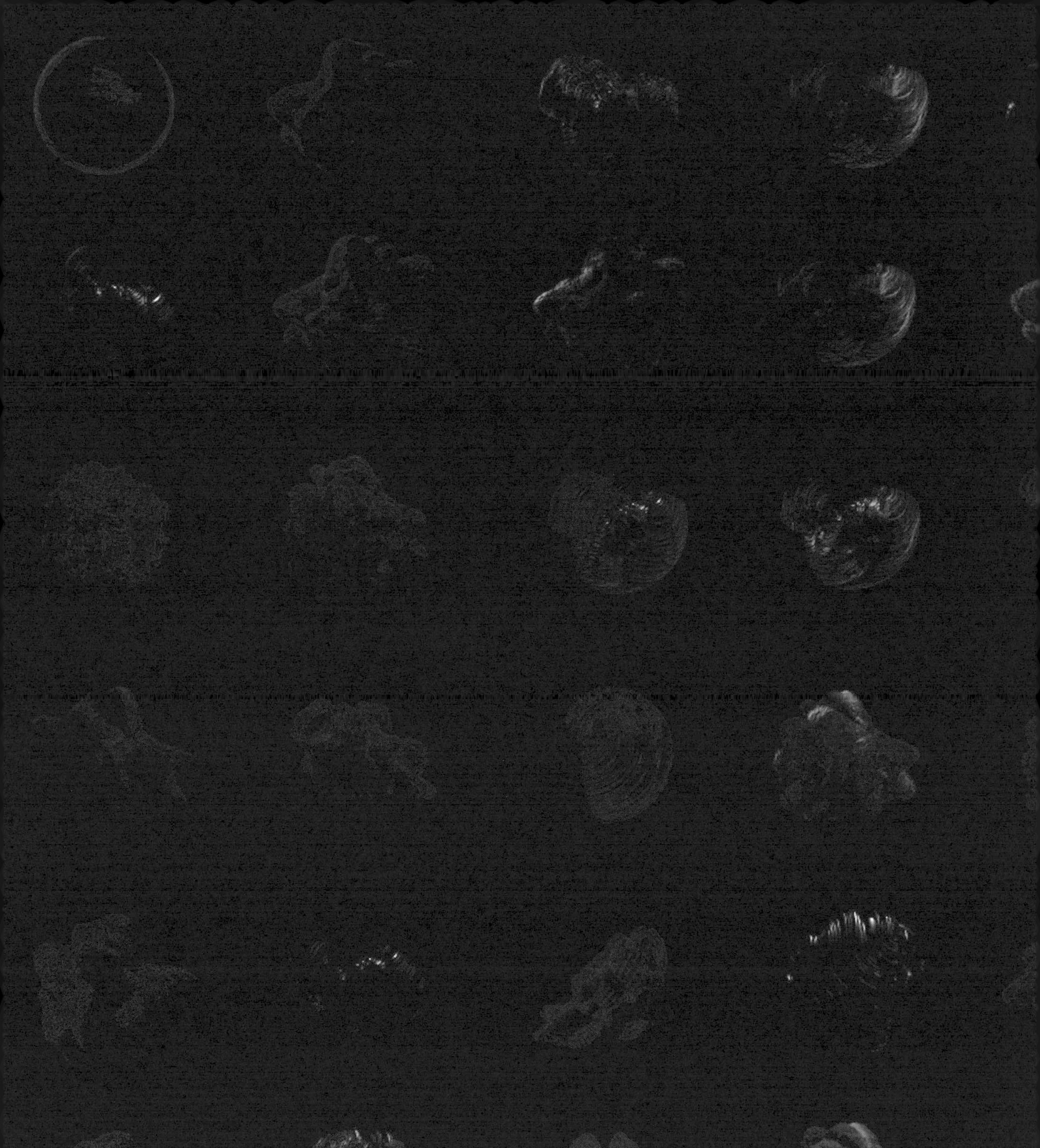

MATERIALISM ITERATIONS. EXCERPT FROM PROGRAMMED MATERIALISM: FICTIONAL TRANSLATIONS EXHIBITION AT 17TH VENICE ARCHITECTURE BIENNALE ITALIAN PAVILION SEZIONE DEL PADIGLIONE ITALIA – CITY X. DANIELA ATENCIO. 2021.

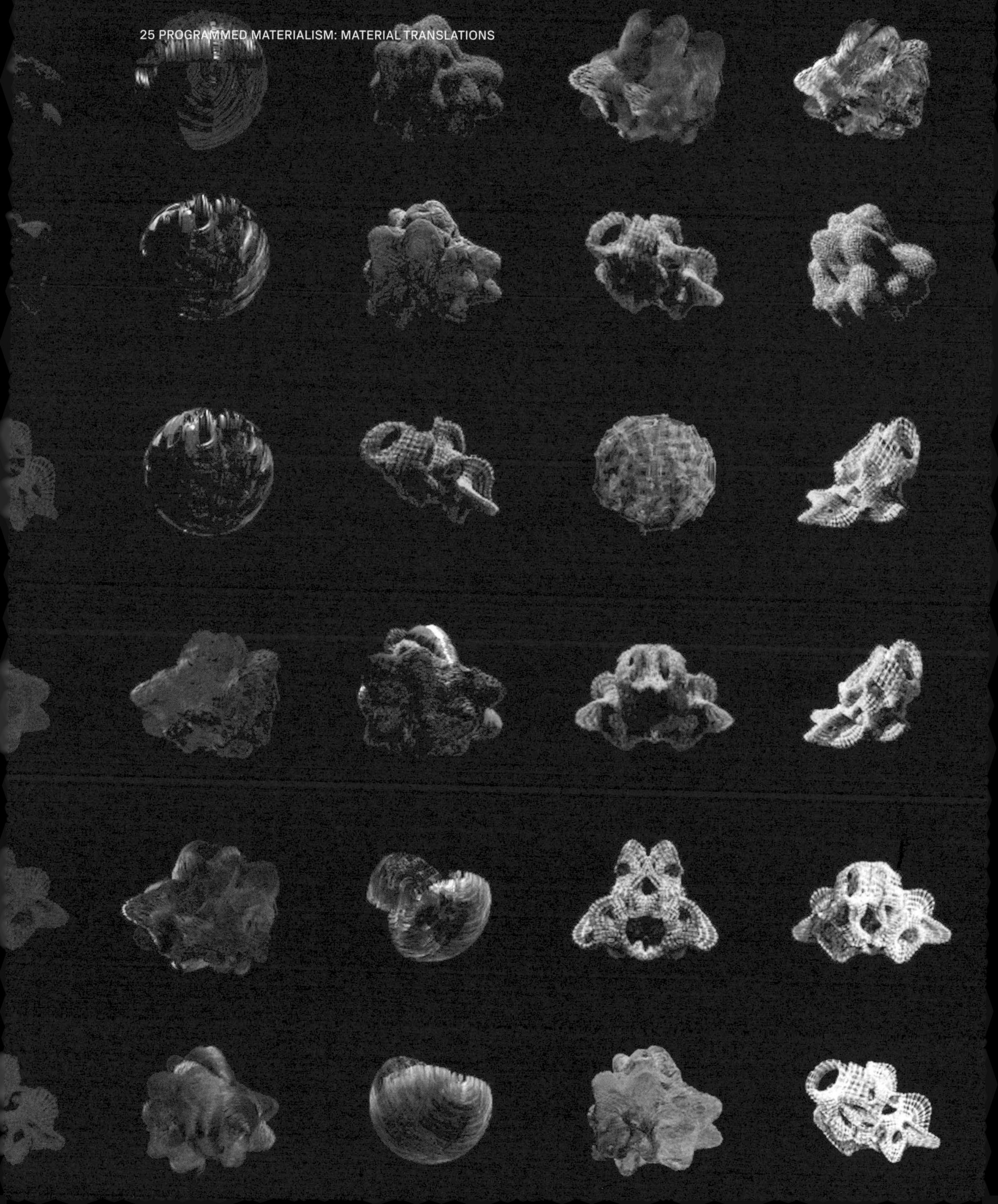

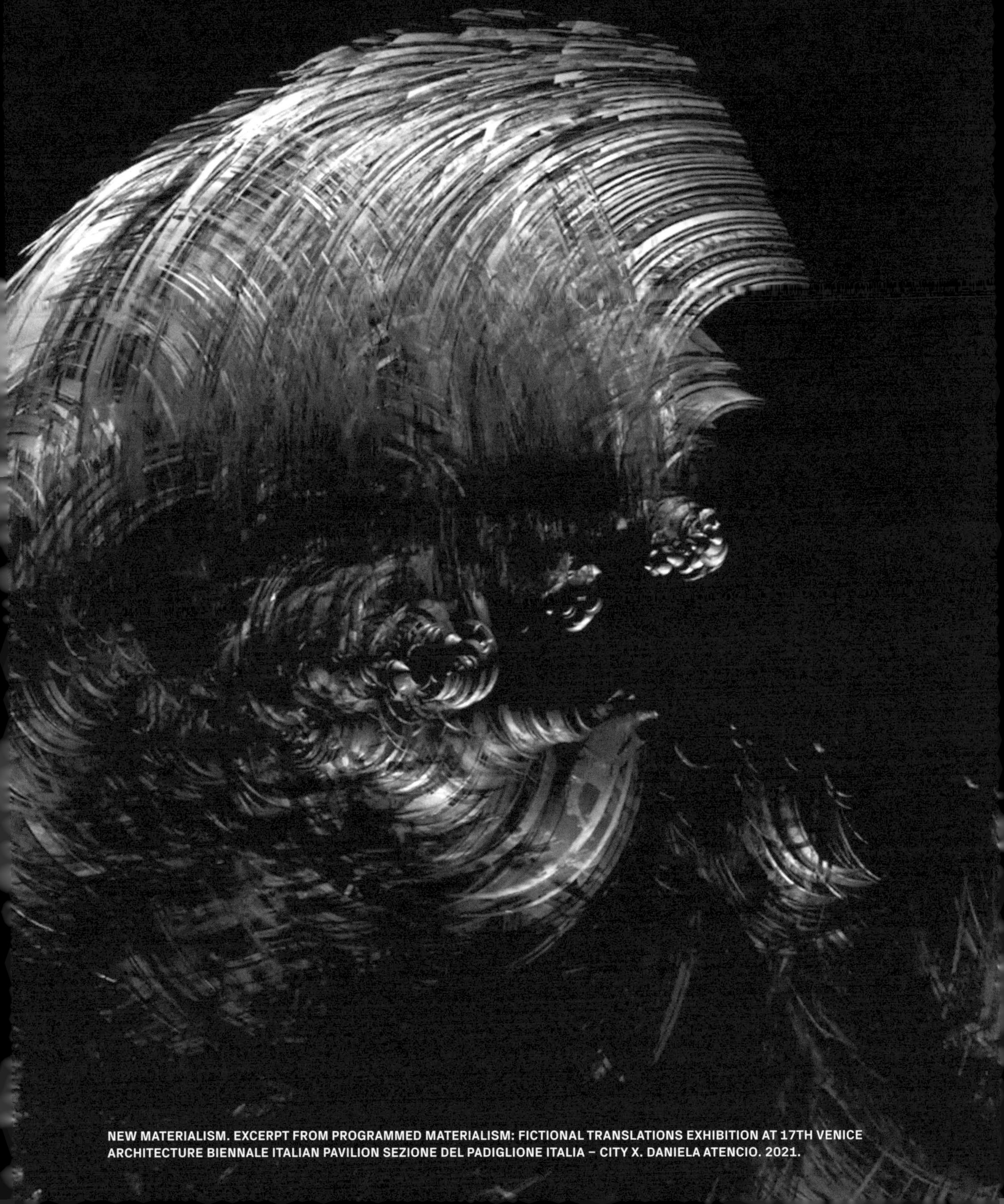

NEW MATERIALISM. EXCERPT FROM PROGRAMMED MATERIALISM: FICTIONAL TRANSLATIONS EXHIBITION AT 17TH VENICE ARCHITECTURE BIENNALE ITALIAN PAVILION SEZIONE DEL PADIGLIONE ITALIA – CITY X. DANIELA ATENCIO. 2021.

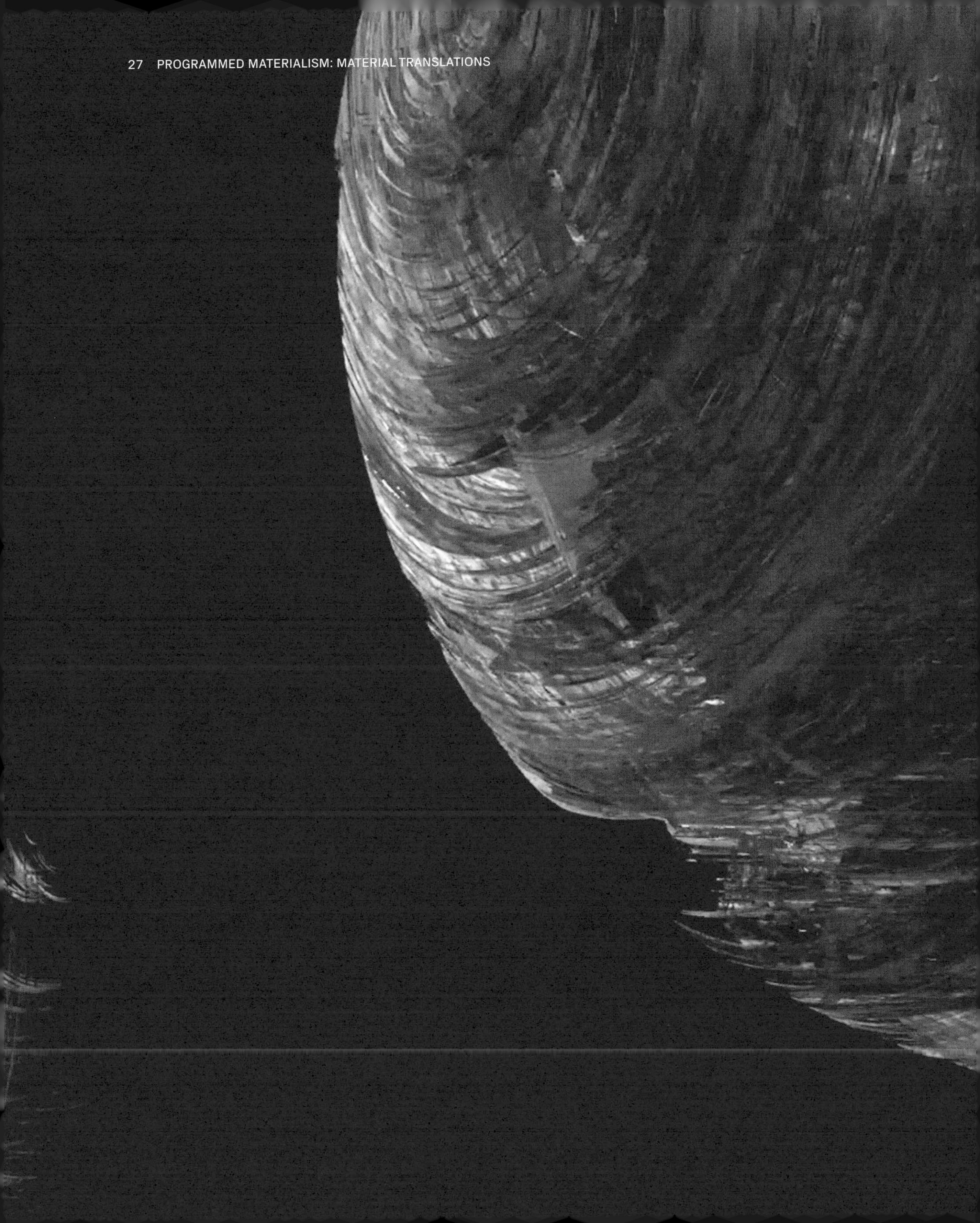

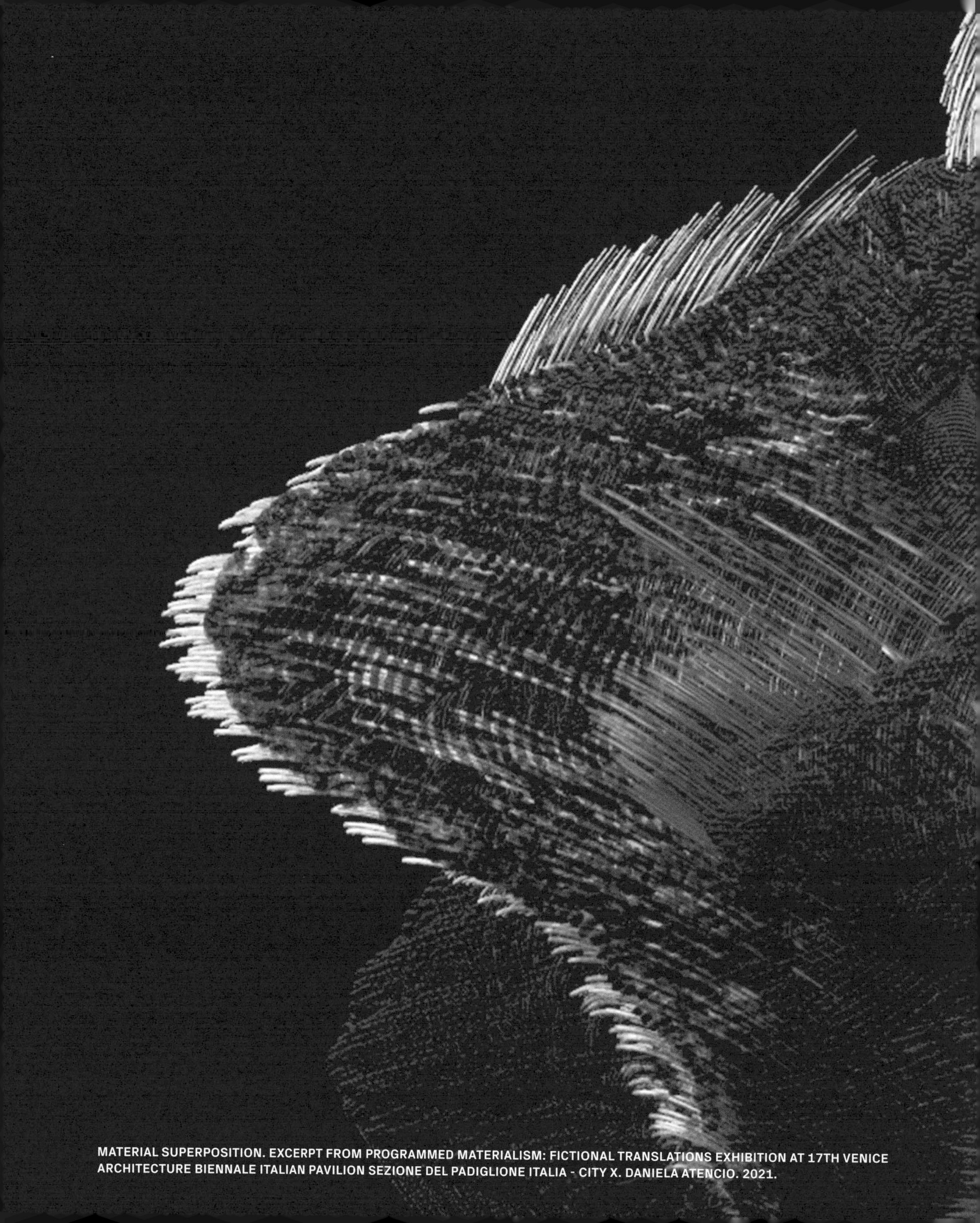

MATERIAL SUPERPOSITION. EXCERPT FROM PROGRAMMED MATERIALISM: FICTIONAL TRANSLATIONS EXHIBITION AT 17TH VENICE ARCHITECTURE BIENNALE ITALIAN PAVILION SEZIONE DEL PADIGLIONE ITALIA - CITY X. DANIELA ATENCIO. 2021.

MATERIAL SUPERPOSITION. EXCERPT FROM PROGRAMMED MATERIALISM: FICTIONAL TRANSLATIONS EXHIBITION AT 17TH VENICE ARCHITECTURE BIENNALE ITALIAN PAVILION SEZIONE DEL PADIGLIONE ITALIA - CITY X. DANIELA ATENCIO, 2021

UNLOCKING THE MATERIALITY. EXCERPT FROM PROGRAMMED MATERIALISM: FICTIONAL TRANSLATIONS EXHIBITION AT 17TH VENICE ARCHITECTURE BIENNALE ITALIAN PAVILION SEZIONE DEL PADIGLIONE ITALIA - CITY X. DANIELA ATENCIO. 2021.

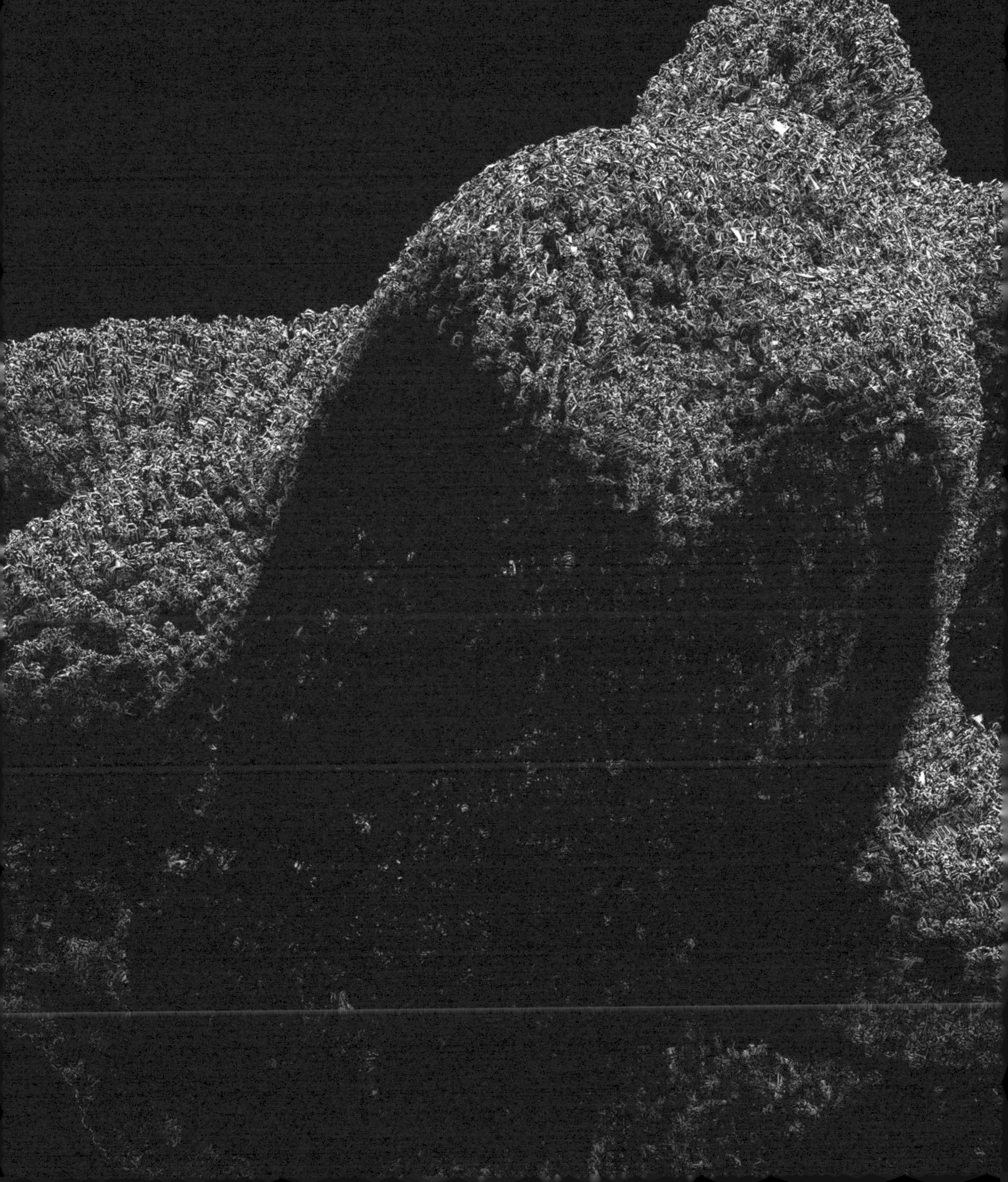

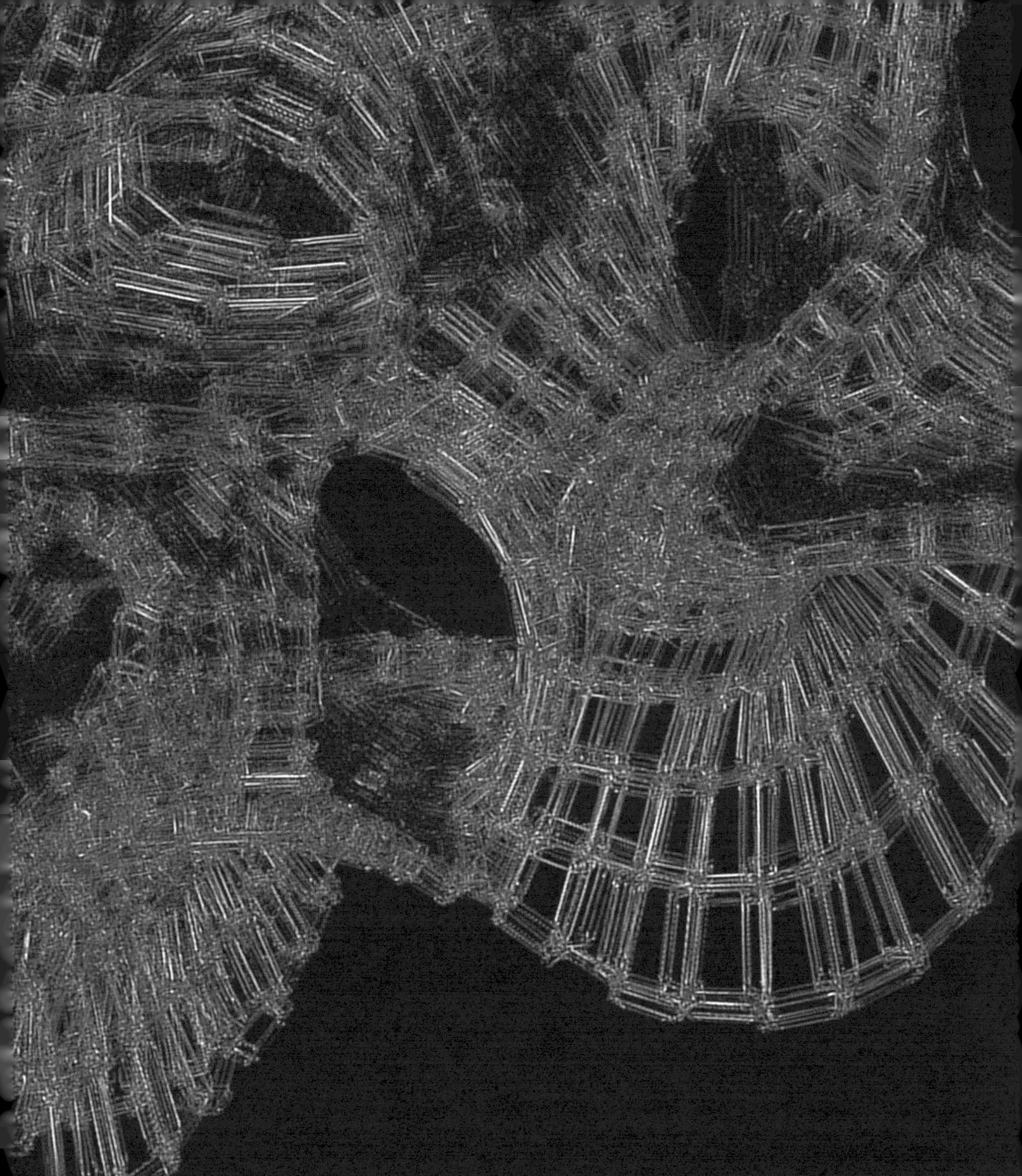

UNLOCKING THE MATERIALITY. EXCERPT FROM PROGRAMMED MATERIALISM: FICTIONAL TRANSLATIONS EXHIBITION AT 17TH VENICE ARCHITECTURE BIENNALE ITALIAN PAVILION SEZIONE DEL PADIGLIONE ITALIA - CITY X. DANIELA ATENCIO. 2021

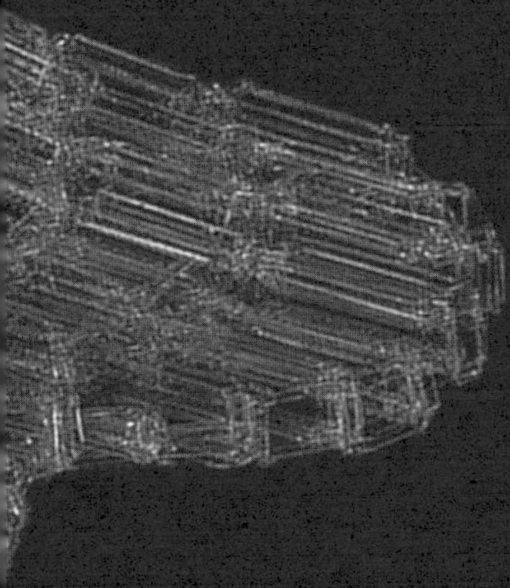

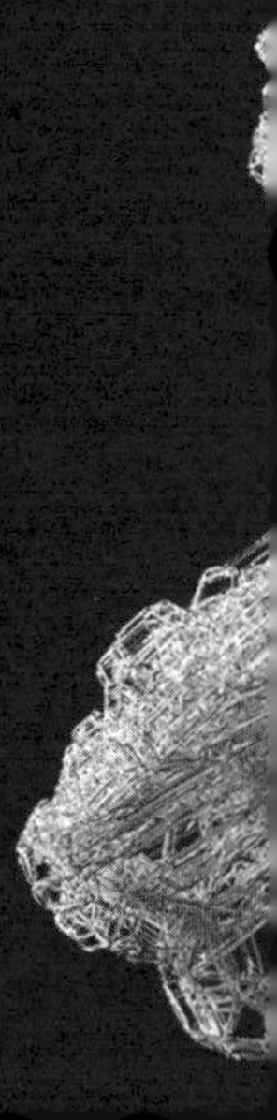

UNLOCKING THE MATERIALITY. EXCERPT FROM PROGRAMMED MATERIALISM: FICTIONAL TRANSLATIONS EXHIBITION AT 17TH VENICE ARCHITECTURE BIENNALE ITALIAN PAVILION SEZIONE DEL PADIGLIONE ITALIA - CITY X. DANIELA ATENCIO. 2021.

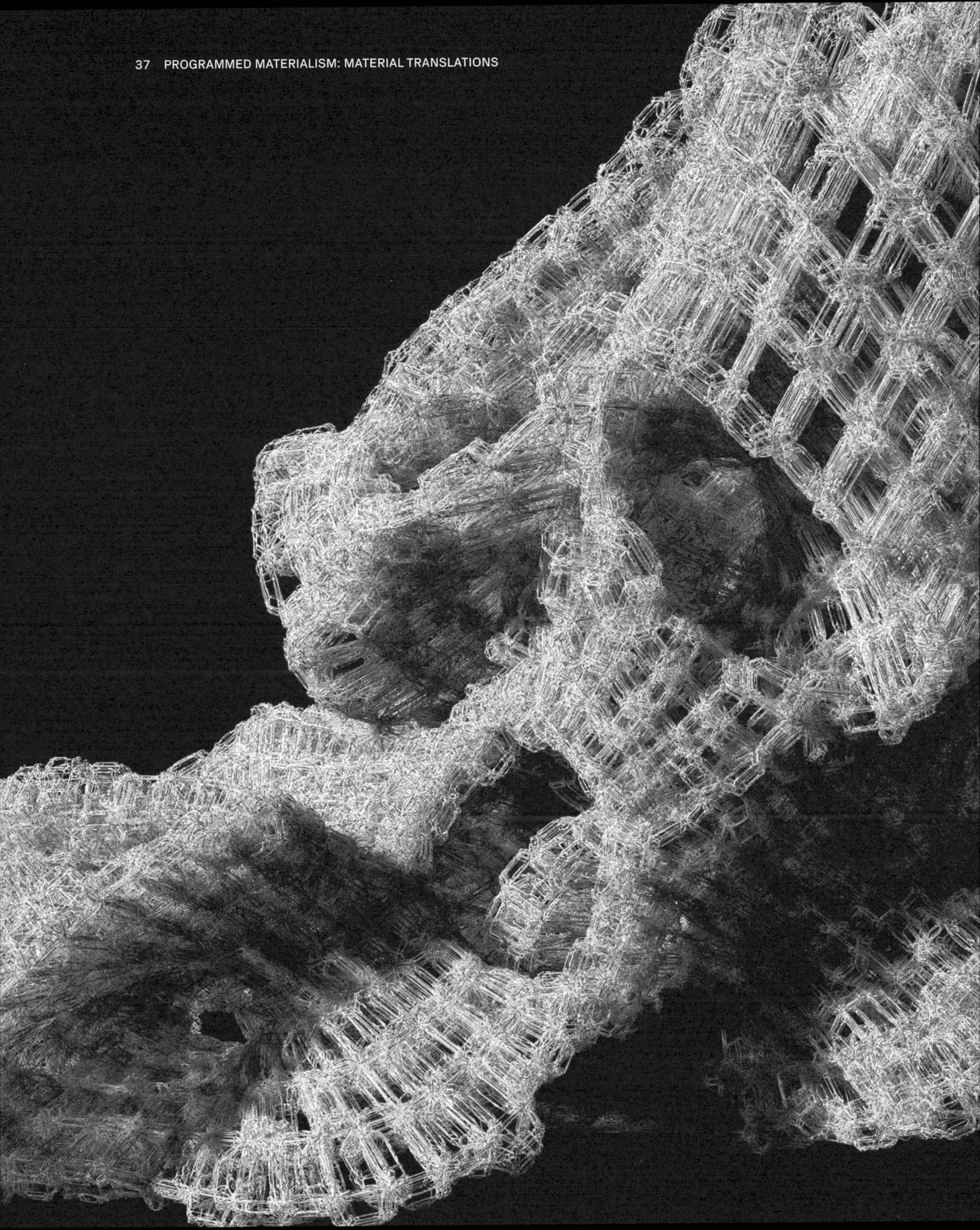

CONTEXTUALIZING DECONTEXTUALIZED OBJECTS. EXCERPT FROM PROGRAMMED MATERIALISM: FICTIONAL TRANSLATIONS EXHIBITION AT 17TH VENICE ARCHITECTURE BIENNALE ITALIAN PAVILION SEZIONE DEL PADIGLIONE ITALIA - CITY X. DANIELA ATENCIO. 2021.

CONTEXTUALIZING DECONTEXTUALIZED OBJECTS. EXCERPT FROM PROGRAMMED MATERIALISM: FICTIONAL TRANSLATIONS EXHIBITION AT 17TH VENICE ARCHITECTURE BIENNALE ITALIAN PAVILION SEZIONE DEL PADIGLIONE ITALIA - CITY X. DANIELA ATENCIO. 2021.

CONTEXTUALIZING DECONTEXTUALIZED OBJECTS. EXCERPT FROM PROGRAMMED MATERIALISM: FICTIONAL TRANSLATIONS EXHIBITION AT 17TH VENICE ARCHITECTURE BIENNALE ITALIAN PAVILION SEZIONE DEL PADIGLIONE ITALIA. [illegible] X. DANIELA ATENCIO, 2021

CONTEXTUALIZING DECONTEXTUALIZED OBJECTS. EXCERPT FROM PROGRAMMED MATERIALISM: FICTIONAL TRANSLATIONS EXHIBITION AT 17TH VENICE ARCHITECTURE BIENNALE ITALIAN PAVILION SEZIONE DEL PADIGLIONE ITALIA - CITY X. DANIELA ATENCIO. 2021.

ROBOTIC CIRCUMSTANCES: DESERT Z EXHIBITION. MATERIAL CLOSE-UP DURING EXHIBITION. 2023

LATIN-AMERICA IN WORM EYE-VIEW

DANIELA ATENCIO
CLAUDIO ROSSI

From Down_Below

In Volume No. 2 of the *Perspectiva Pictorum et Architectorum*[1] treatise, Jesuit Father *Andrea Pozzo* writes to the reader warning about the shift in the orientation of the gaze from perspective to "*reorder the world and submit it to reason*"[2] Later, the humanists of the fifteenth century positioned the use of perspective as a tool for understanding and reorienting the world. In this sense, perspective is a scientific method at the service of three-dimensional central vision and not a tool for the creation of a false spatial illusionism. The viewer gaze merge with the target point to generate an authentic and accurate knowledge of reality. In other words, the representation of the observation becomes a source of knowledge.

Pozzo exemplifies that rational and scientific observation in the use of perspective to look from below and above in the project for the dome of *Sant'Ignazio di Loyola* in Rome[3]. The target point of view is the elevated gaze where all the lines are projected 'leading' (or observing) God. This 'humble' look from below to something superior and sublime is where the position of the *Worm-Eye View* [*W-EV*] is located.

The *W-EV* is presented as a counterpoint to the lofty, grandiloquent, territorial, and dominant view from above. Already during World War II, from the dominance of the air exercised by aircraft squadrons, the so-called *Bird-Eye View* contributed to territorialize the gaze on geography[4]. The maps used to conquer territories arise, as *Stuart Elden* explains, as a form of "aerial terror", so terror and territory became etymologically linked.[5] *R.F. Delderfield* constructed theatrical scripts, later taken to the cinema, which narrate the experiences of those who in their daily lives observed the aerial domain from the ground during World War II, referring to them as "worms"[6].

One of the first representations and descriptions of an upper-grandiloquent view in our region appear in the first botanical expeditions of the early 18th century and, in particular, in *Alexander von Humboldt's* description of the Amazon rainforest when he ascended the Chimborazo in 1802.[7]

In the late nineteenth and early twentieth centuries, illustration, through its satire, succeeded in developing various representational devices that offered a view from below. *Henry Mayer's* drawings, entitled *Worm-Eye Views of Us*, presented a series of views of an everyday-life understood from below and above (Down-Upside). Other illustrators, such as *Roland Hill*, also sought to highlight views through such representations. In these examples, the counterposition to the gaze is presented as different critical scenes making opposition to the representation of 'seeing from above' or 'seeing upwards,' in other words, seeing reality as 'gods.'

In 1945, the painter *Joaquín Cortés-García* created his famous drawing *América Invertida*, in which he proposed a shift in the perception of our context, coining the phrase "*our north is the south*"[8]. By observing from below and turning the perspective on our region, a paradigm shift is produced that insists on a more complete view of our reality. A kind of *W-EV* on Latin America.

In the twentieth century and especially remembering *Pozzo*'s efforts, many exercises of oblique representations came to architecture and many of them in *W-EV* axonometries.

1 *Pozzo*, A. (1693) "*Perspectiva Pictorum et Architectorum*". Rome: Typis Joannis Jacobi Komarek.

2 *ibidem*

3 *ibidem*

4 Elden, S. (2009) '*Terror and Territory*'. Minneapolis: Minnesota University Press.

5 *Ibidem*. In addition, to expand the definition of territory (and territorial), see also: Elden, S. (2013) "*The Birth of Territory*". Chicago: Chicago University Press.

6 Delderfield, R. F. (1945) "*Worm-Eye View*". Theater play.

7 Wulf, A. (2016) "*La Invención de la Naturaleza. El Nuevo Mundo de Alexander von Humboldt*". Madrid: Taurus.

8 Although this phrase was propagated as part of the artist's *oeuvre*, it is noted both in the work and in many formal and informal interviews. This phrase drives concrete works of literature such as *Mario Benedetti's* poems, "*El Sur también Existe*" (1986). These poems are connected to the homonymous song by *Joan Manuel Serrat*.

LATIN-AMERICA DRAWING IN WORM-EYE VIEW: EXPLODED LAYERS FROM DIGITAL TRANSLATIONS PROCESS. MESHES, POINT CLOUDS, AND DENSITIES.

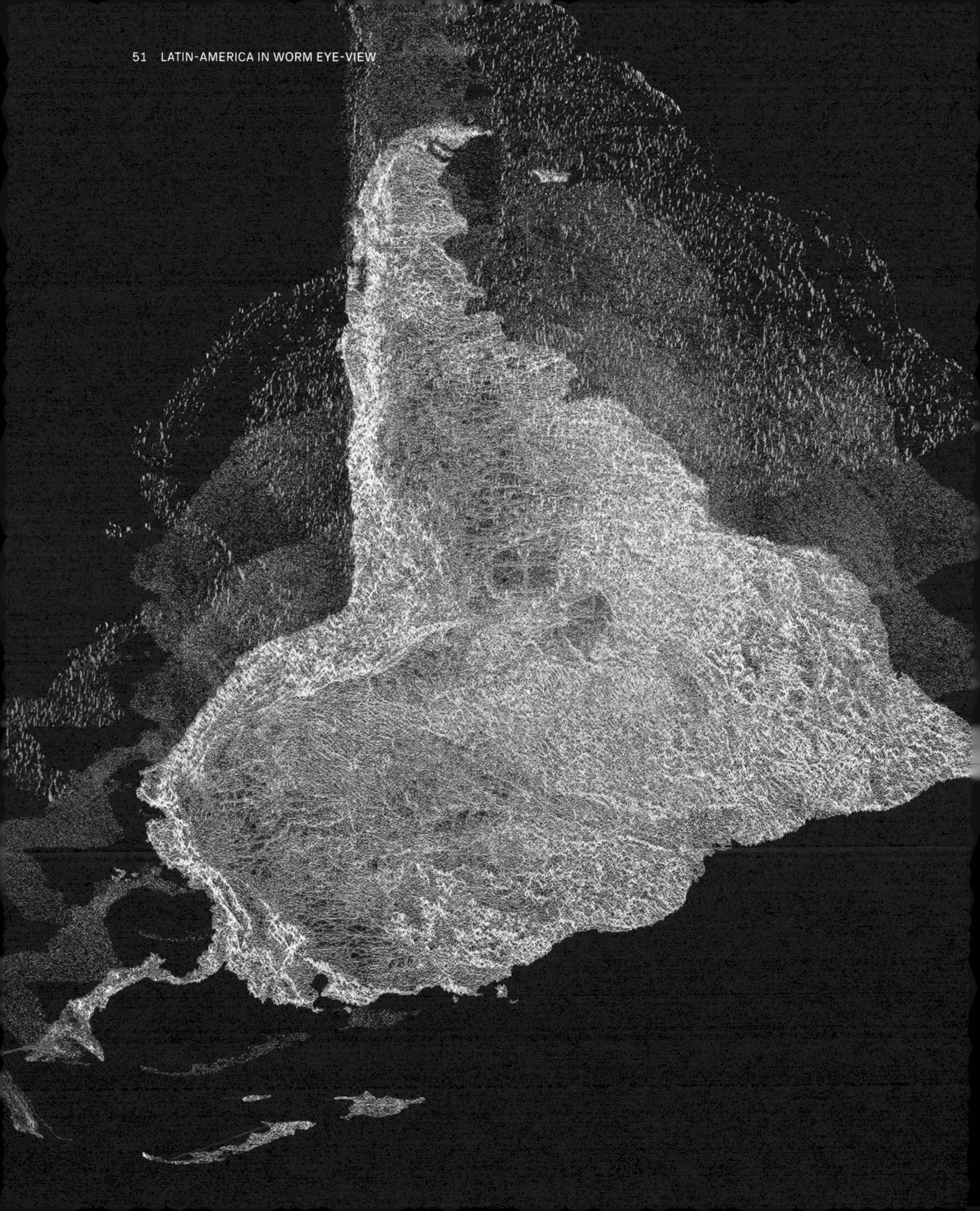

Mario Botta and *Rafael Moneo*, among others, conceived architectural projects from this positional observation of their buildings. The architectural plan as a two-dimensional drawing (and seen from above) would not cover the architectural capabilities of the space explored in the design exercise.

Architect *Enrique Ciriani* has insisted on the need to think the habitable space contained between the obverse and reverse of the architectural ground[9] (*Surface*[10] in French). Space then is a habitable continent between the two sides of the surface. The philosopher *Martin Heidegger* deeply develops this reflection from the phenomenological condition of inhabiting in *Bauen, Wohnen, Denken*, when he says: "(...) *To inhabit is to 'be'*[11] *between the ground and the* sky"[12] If we take up *Ciriani*'s reflections, from this phenomenological definition, we can say that representation, from the observation of the habitable space, must reorder hierarchies and shape new approximations from the architectural exploration. An architecture experience deployed in a spatial thickness delimited by the sky,[13] the reverse, and the ground, the obverse.

We imagine, as an initial approach to this text, an epistemic repositioning around representation, but above all (or rather, to be consistent with the argument), a change in the gaze of habitable space in Latin America and its form of representation. *W-EV* observations that allow, from the ground, to make an expedition to the sky. From below.

Magical-Realistic' Speculative Tools

When *Cortés-García* coined his phrase, he established a point of view and an observation. The point of view, with political and ideological undertones, focuses the gaze on Latin American territory as a platform for exploration and speculation. If the north is the south, then this must be the place from where our research questions and the thrust of innovation must be produced. In any case, looking at Latin America as a place is in itself a valuable reflective exercise in seeking unique ways of working.

All creative fields, especially architecture, must be aware of the Latin American reality. This awareness must be translated into the promotion of workspaces that blur disciplinary paradigms and favor the intersubjective construction of our own reality. Reality itself, with its complexity and challenges, but also with its optimistic potential, is the principle that governs this observation proposed by the Uruguayan artist Cortés-García.

García Márquez, *Uslar-Pietri*, *Benedetti*, *Borges*, *Watanabe*, and other great Latin American writers have developed their novels from the re-observation of the Latin-American reality. *García Márquez* for example, once said: *"The first condition of magical realism, as its name implies, is that it is a strictly fact that, nevertheless, seems fantastic (...) in Latin-America there is no greater fiction than reality itself!."*[14] The description of this fiction is the platform that can allow rethink local disciplinary positions, operations, and representations of the reality.

For this reason, the contemporary architectural project must establish new agencies with the context in which it is implanted. The key to the contextual question lies in the deep understanding and reinterpretation of the multiple conditions

9 See: Ciriani, E. (2014) "*Todavía la Arquitectura*". Lima: Editorial Arcadia.

10 According to *L'Académie Française: 'Surface'* is: "*n. f. Superficie, partie extérieure d'un corps, ce qui le limite dans l'espace. Surface plate, unie, raboteuse. La surface de la terre. La surface de l'eau. Une surface plane. Une surface concave. Une surface convexe. Mesures de Surface. Surface signifie au figuré Extérieur, dehors, apparence*". Highlighted by the authors.

11 '*Being*' in this case refers to the *Heidegger's* phenomenological *Dassein* or '*being in*'. See: Heidegger, M. (1956) "*Construir, Habitar, Pensar*". Trad. (2015) Jesús Adrián Escudero. Barcelona: Oficina de Arte y Ediciones.

12 *Ibidem*. Translated by the authors.

13 The similarities that these expressions have with terms used in the architectural profession such as '*soffit*', which comes from the Latin '*suffixus*' or 'fastened below', or in Spanish 'cielo raso', constitute a legacy of approximations that seem disconnected but are really contained in the most important architectural constitution: space.

14 García Márquez, G. (2000). Interview for the newspaper La Reforma de México.

in which the architectural design exploration operates. Reality is determined by superimposed layers of complexity and often subtle temporal processes that go beyond simple observation. This is where the technological tools and the representation play a fundamental role in the speculative possibilities of the project. It is at this point where the representation of a context from a different perspective and the construction of an apparently illusionistic reality must summon new (or recover old) observation conceptualizations.

The world, and Latin America in particular, in its most recent state of convulsion, becomes a space to be speculated and observed with other 'eyes' and with other 'lenses'. The etymology of '*speculum*,' which originates from the archaic Latin '*specere*' or 'to look,' closely relates to the act of speculation. The path must then be speculative—to mix ways of seeing and ways of representing the observed reality. The *W-EV* will be the form, from the down-upside gaze, and the technology, especially the robotic arm, the way to propel this gaze in the selected cases to investigate the region.

The robotic arm and especially all the pedagogy built around it, leads to a series of exercises of speculation and exploration based on methodologies such as Research Through Design or Learn by Doing. This pedagogy has promoted exercises of approximation to the facts, phenomena and cases that are far from being conventional and that on the contrary reinforce the change on certain observations. If the tool changes its approach to design, then the exercise of the representation must be consistently different. If the representation changes —in its way of describing reality, representing it and doing so, the tools and methods must be consistent with this assertion. This is why the space of Robot-Lab[15] was born, as a laboratory to change the look through the use of a tool as special for our context as the robotic arm.

The cases presented in this book were selected with precision and delimited by a series of fundamental aspects in the *Programmed Materialism* research.[16] On the one hand, to look for spaces where history, materiality and landscape collide in such way that the capacities of the explorations could generate design processes that, could contribute to the local logics and the particular needs of these contexts. In other words, where everyday life and its reality would inform the different project inquiries; and on the other hand, that could contribute to the general reflection on the Latin American regional platform as a potential space for the resolution of local environmental or social conflicts through design.

To approach this intricate complexity and address a possible triad (history, materiality, and landscape, without order or hierarchies) two assessments are established: The first, from *Fernand Braudel* temporal conceptualization, not only in his layers of time analysis (short, medium and long duration) but also with his rigorous historiographic analytical tools: diachrony and synchrony.[17] The second, from the use of technology as a design vehicle an a working tool work in each presented process.

This generates a set of tools with which, like surgeons, it is possible to observe and operate on the 'body' of the selected sites and make project decisions. Continuing with the anatomical metaphor —and recalling *Alberti's* analogies—, technology made possible to address not only aspects of a geographical scale, but also material micro-sites in the effects of time (e.g. erosion) of the ruins found. This means looking closely at

15 Robot-Lab is the undergraduate architecture project workshop at Universidad de Los Andes, presented in other parts of this book.

16 This is the name of Professor Daniela Atencio's research-creation project. Much of the results shown in this book are part of this exercise.

17 Braudel, F. (1958) "*History and the Social Sciences: The Longue Durée*". Review Journal, Research Foundation of State University of New York, 32 (2) 2009. To expand these ideas about the historical time definition, see also: Braudel, F. (1982) "Civilization and Capitalism, 15th-18th Century, Volume I: The Structures of Everyday Life". Berkeley, University of California Press

he body and its materiality, or the 'meat' (flesh) as lberti named in his treatises.[18]

In part, it is argued from different histori-al positions because the selected cases involve he accumulation of temporal layers, such as the valled city of Cartagena de Indias in Colombia. his military artifact, built between 1614-1796, is he perfect synthesis of the issues addressed in his reflection. There are many things about it that nade it interesting for this research: its history, its ossible environmental value as a physical barri-r against a Caribbean Sea that is losing species nd being affected by the planet's environmental rama due to rising temperatures (e.g.: *sea level ise*), its social and cultural significance, and the egative connotations of not caring about its sur-oundings and dividing social classes, are just a ew of them.

Another case study faces the same reality. he Guajira's desert, on the northernmost coast f the southern shelf of the American continent, hares a regional extension between Venezuela nd Colombia. The political-administrative divi-ion between both countries does not condition eparations in the understanding of this histori-al region.[19] In fact, it is the wonderful transitive nd expansive condition that promotes a desert etween two countries that makes similar environ-nental, cultural, and social realities, to generate ery specific design processes, but also replicable r extensive.

In other words, one of the cases is situat-ed, static, stationary, and expectant. The other is xpansive, permeable, and diffuse. One is a wall, a ertical surface possessing thickness, *thingness,* nd programmatic capabilities. The other is a 'car-et' that supports everything, a horizontal surface that likewise accumulates thicknesses and con-tents between the sky and the arid ground.

W-EV Procedure: *t+r*

Like every process presented in this book and with the objective of constructing pedago-gies associated with the architectural project, we end this chapter with a procedure. In this case, the proposed procedure is related to the condition of observation and the challenges associated with a position from and for Latin America.

This procedure seeks to establish links between the questions and the studied cases presented previously. In this sense, the two main considerations presented as working questions for this procedure are: time (*t*) and responsibili-ties (*r*). We have emphasized the need to adopt a contemporary gaze in each of our architectur-al project-related workspaces, acknowledging that traditional working conditions, like the Vit-ruvian triad or the recurring dichotomies of craft, like the form or function of the architectural pro-ject, are insufficient. On the contrary, in order to address the marvelous complexity of Latin Ameri-can reality, it is necessary to ask questions about temporality and the specific responsibilities of the project in the context. To shift the observation.

Time (*t*) is conceived as the first of the issues to be represented from the *W-EV*. From *Brau-del's* perspective, time is the exercise of thinking as a diachronic and synchronic structure, which must act both in a superficial layer and in the long duration. However, this reflection is not enough to achieve a greater meaning or to weight time within the projectual engine. The building as an artifact[20] entails, at least, a reflection on useful, materi-al, and programmatic time. Questions about this idea derive then towards temporality as a prop-

18 Alberti, L.B. (1485) "*De re ædificatoria: Vols. VI–XIX*". Translated by Rykwert, J., Tavernor, R. & Leach, N. (1988), Cambridge: MIT Press.

19 Galué, G. (1998) "*La Región en el Proceso Histórico Venezolano. Propuesta de Periodización para la Región de Maracaibo*". Caravelle. Cahiers du Monde Hispanique et Luso-Brésilien, 70. See also: Cardozo Galué, G. (2013) "*Significaciones y Alcances del Concepto Región Histórica*". Cambios & Permanencias (4): 45 – 61.

20 In the different parts of this book there are transversal and connective references, which replace the conventional building definition with the 'artefact' or 'architectural artefact', to propose then a Hyper-Artifact. This refers back to the same conversation about architecture as infrastructure and machines, that represent and are represented through the use of technology.

erty of the artifact, inherent to its own projectual genetics. What chronological processes are contemplated? What are its material capacities? Is it about ruins? Is the artifact flexible, ever-changing, adaptive? Does it grow or disappear? Does it involve a reflection on climatic or ecological time? There are many possible questions that derive from this section and that nourish the reflection of the project in each speculative exercise.

If the previous questions announce certain reflections on the capacities of the architectural project, it is from the responsibilities (*r*) that we can spatialize them giving them thicknesses. By approaching the project through the lens of responsibilities, we can generate remarkable insights into buildings as artifacts situated within specific contexts. That adds, restores, and contributes to the particular logic of the contexts in which they operate. Again, it is not enough to think, for example, of a school for a specific municipality, only because of the lack of public services that this hypothetical context possesses. From the position presented in this procedure, the school must also be an artifact that retains rainwater or generates energy and also serves as a possible emergency shelter.

These should be artifacts to, in the Latin American case, turn our gaze and think of ourselves in a different way, giving it more capacities—that is to say, more responsibilities. Multiple deficits need to be addressed, with an incredible potential in our own logic and capabilities. Issues such as sustainability are just the tip of the iceberg of a position that is based on time but also on the responsibilities of the project.

Any similarity with other realities of the Global South is not a mere coincidence. Please always refer to the equation:

WE-V=t+r

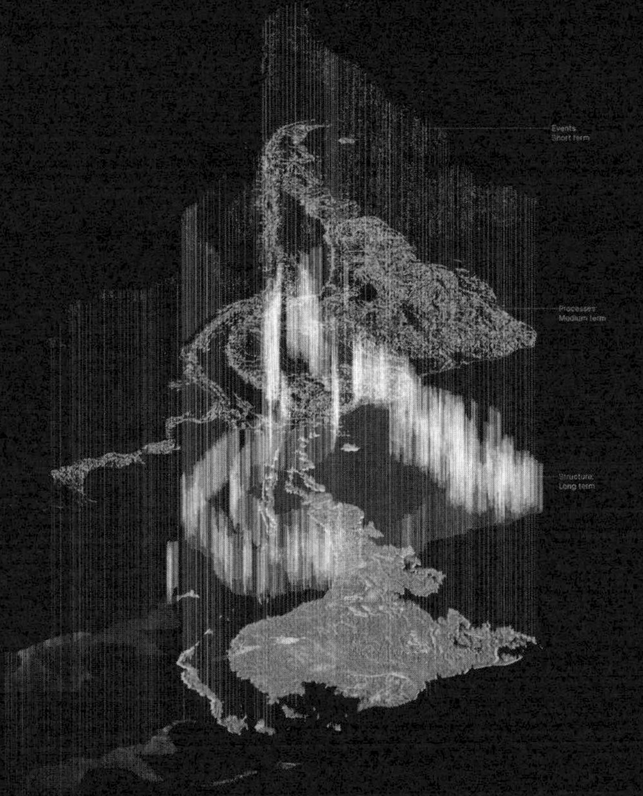

LATIN-AMERICA DRAWING IN WORM-EYE VIEW: EXPLODED LAYERS. DIAGRAMMATIC INTERPRETATION OF THE STRUCTURAL TIMES OF FERNAND BRAUDEL'S SOCIETY, ACCORDING TO CIVILIZATION AND CAPITALISM, 15TH-18TH CENTURY, VOLUME I: THE STRUCTURES OF EVERYDAY LIFE, Y HISTORY AND THE SOCIAL SCIENCES: THE LONGUE DURÉE.

TRANSLATIONS: METHODOLOGIES

DANIELA ATENCIO

Translations, in order to establish what the title of this book proposes, focuses on the encounters and transits between different worlds. From the analogue to the digital and post-digital; from the physical or 'real' back to the digital[1], or from the human and the non-human. Contrasting two apparently opposing alternatives but which, with the necessary bridges, made it possible to build new worlds from their correlation. We speak of bridges, from its own definition of infrastructure, because of the capacity to connect and establish links between two opposing boundaries, overcoming any obstacle.

The interesting thing about this metaphor is precisely the capacity that exists in a bridge to join two extremes (*Martin Heidegger's* 'here' and 'there'[2]) but also the capacity to let transit, in both ways, the apparatus that joins them (remembering again *Heidegger's* 'in'[3]). Through these technological bridges we travel at different and differentiated speeds, but sometimes that speed slowed, becoming zero, to stay and to remain on it. It was a movement of coming and going, a becoming, feeding, and feeding back into the various design processes presented here.

Sonia Colina defines *translations* as "*an activity, and a product*"[4], in a transit that is often difficult to determine because of their overlapping. For this reason, she also states that *translations* can refer to: "*the process that produces the product (also referred to as translating) and the product itself (translation or translations)*"[5]. In essence, *translations* in its own technical and linguistic structure generates resulting products as an outcome, but perhaps more importantly, a process. For the purposes of this research-creation book, these two outcomes have merged in the *design processes*.

Some authors involved in the study of *translation* as a field of knowledge insist on the technically stricter basis of language use as distinct from interpretations. An interpretation involves changes and adjustments to the message (language) to adapt it to the receiver's (context). In this case and far from the possible closeness in definition, we used *translations* as a mechanism that involved a specific working process (activity, or in this case, procedures) and the handling of *translating* results (objects) between the different parts of those mechanisms. In other words, the interest is in that bridge between images, codes, fabrications, and assemblages, terms that will be part of the workflow proposed in the following chapters.

The beginning of these *design processes* referred to the image as a capture of the initial reality of a *hyper-artefact*[6]. The robot was the central agent, the bridge, and the tool of *translation* between the digital and the physical worlds.

The aesthetic and readable notions from an image were challenged in representational and robotic fabrication, developing *translations* to speculate on the generation of retracted, distinct, and informed design processes through both algorithmic programming processes, advanced digital fabrication (robotic) and post-computational techniques. The possibilities of *translating* information from images to geometries — and vice versa—activated the different speculative mechanisms and design results presented in the following parts of this book.

Instead of being attached to replicate or re-sampling traditional architectural design methodologies, these *translations* propose a diverse and strange set of mechanisms in the relation between ideas and design explorations, challenging the traditional notions of form finding and representation.

1 This exchanges between the digital and the physical world recalls to Negroponte, N. (1995) "*Being Digital*". New York: Knopf Doubleday Publishing Group.

2 Heidegger, M. (2016) "*Construir, Habitar, Pensar*". Teoría, n.º 5-6 (junio): Pág. 150-162

3 *Ibidem.* The term 'in' is here translated and contracted from Martin Heidegger's '*dassein*' or '*being in*'.

4 Colina, S. (2015) "*Fundamentals of Translations*". Cambridge: Cambridge University Press.

5 *ibidem*

6 See *Hyper-Artifacts* chapter.

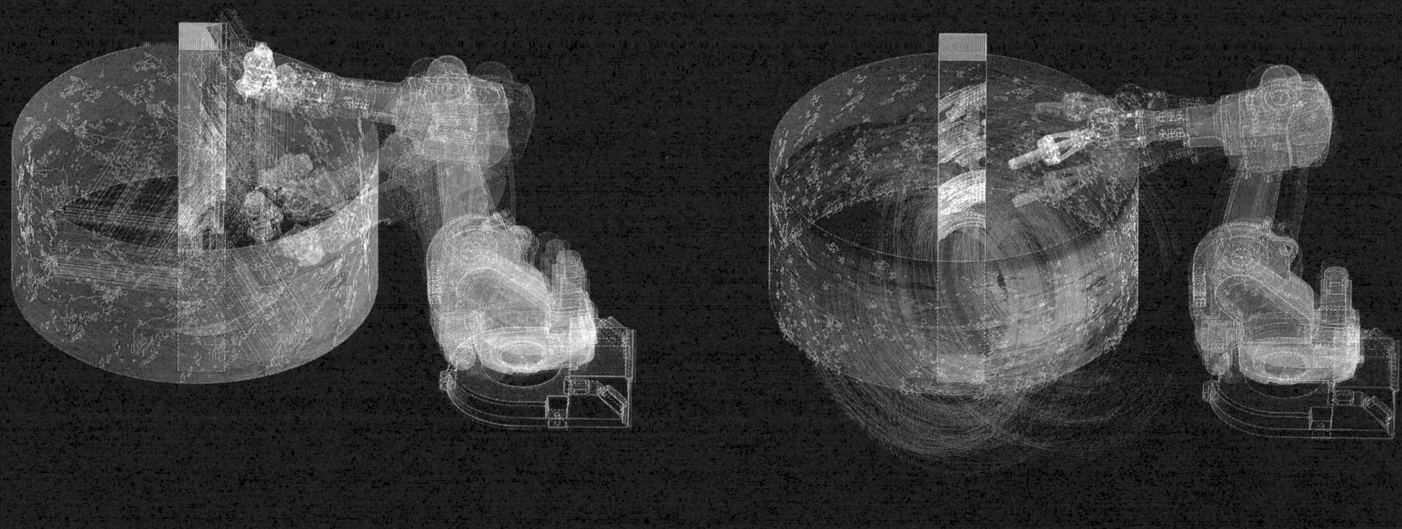

OBLIQUE DRAWINGS: ROBOTIC TRANSLATIONS BY IMAGE, TOOLS, AND CHOREOGRAPHIES.

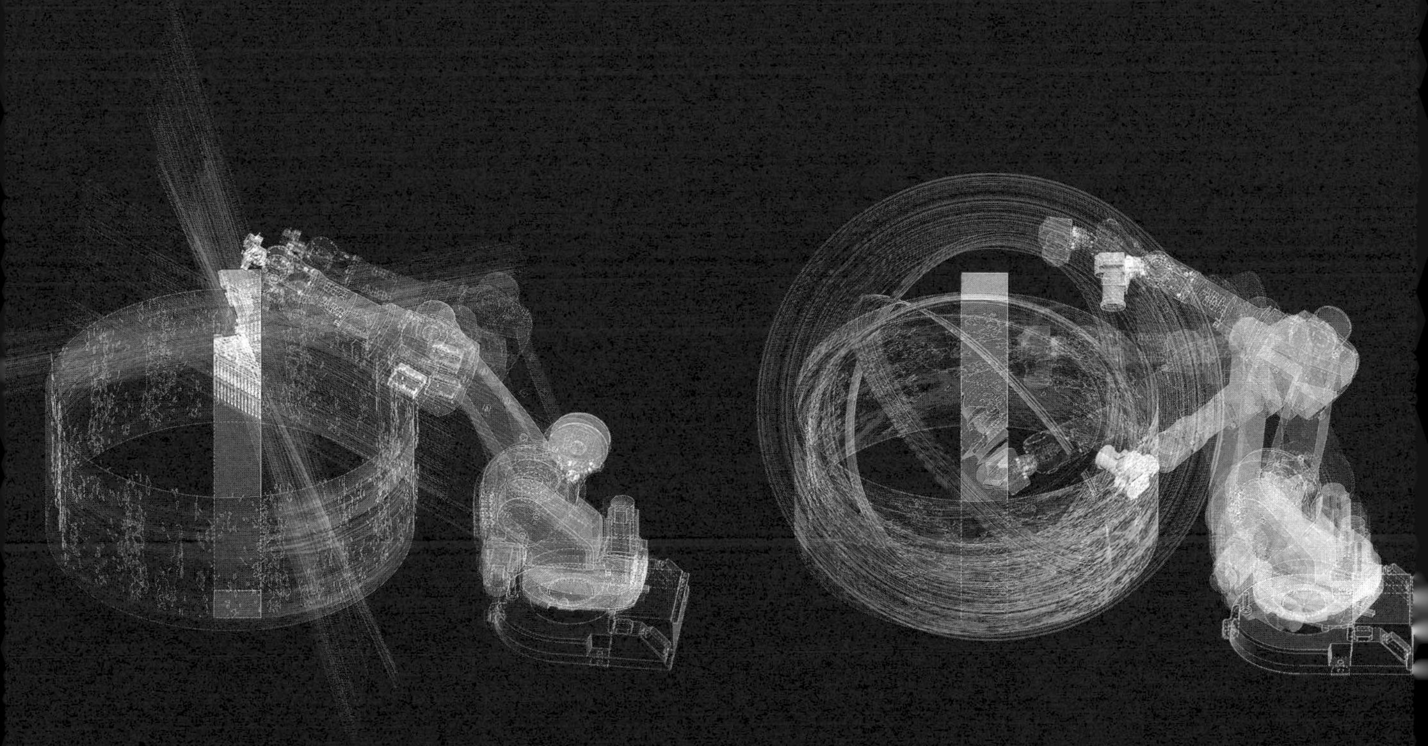

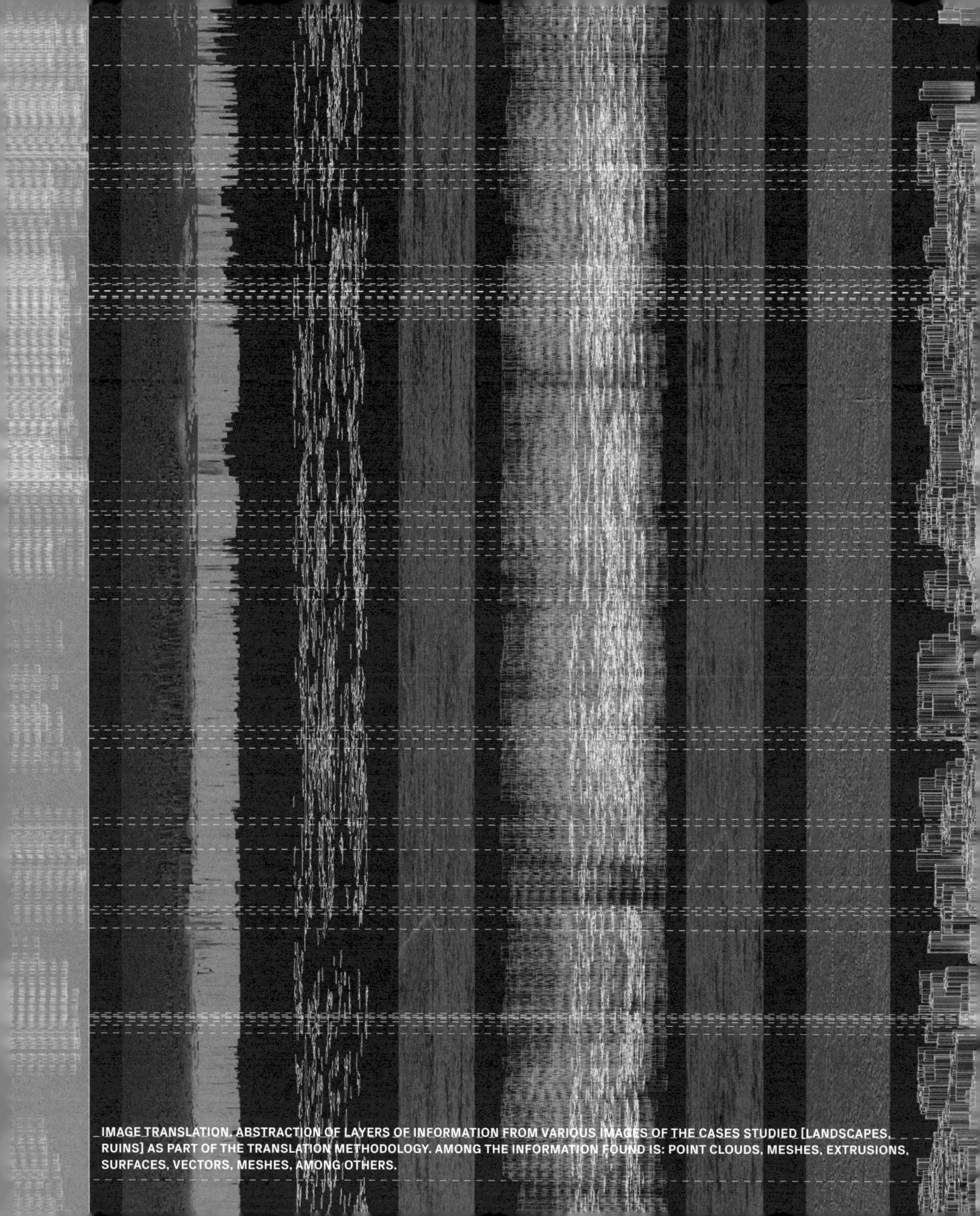

IMAGE TRANSLATION. ABSTRACTION OF LAYERS OF INFORMATION FROM VARIOUS IMAGES OF THE CASES STUDIED [LANDSCAPES, RUINS] AS PART OF THE TRANSLATION METHODOLOGY. AMONG THE INFORMATION FOUND IS: POINT CLOUDS, MESHES, EXTRUSIONS, SURFACES, VECTORS, MESHES, AMONG OTHERS.

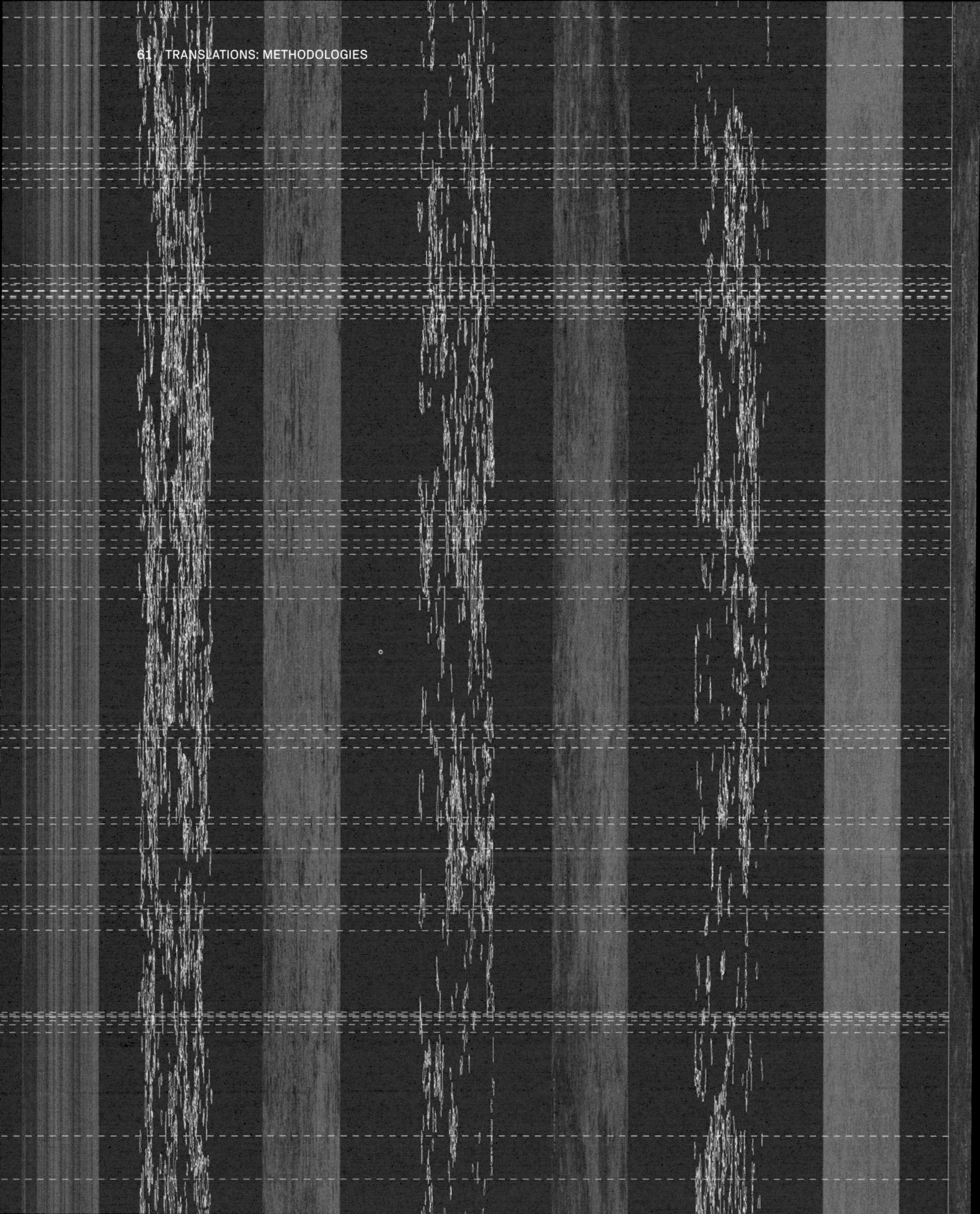

THE ARTIFACT: THE HYPER-ARTIFACTS

DANIELA ATENCIO
CLAUDIO ROSSI

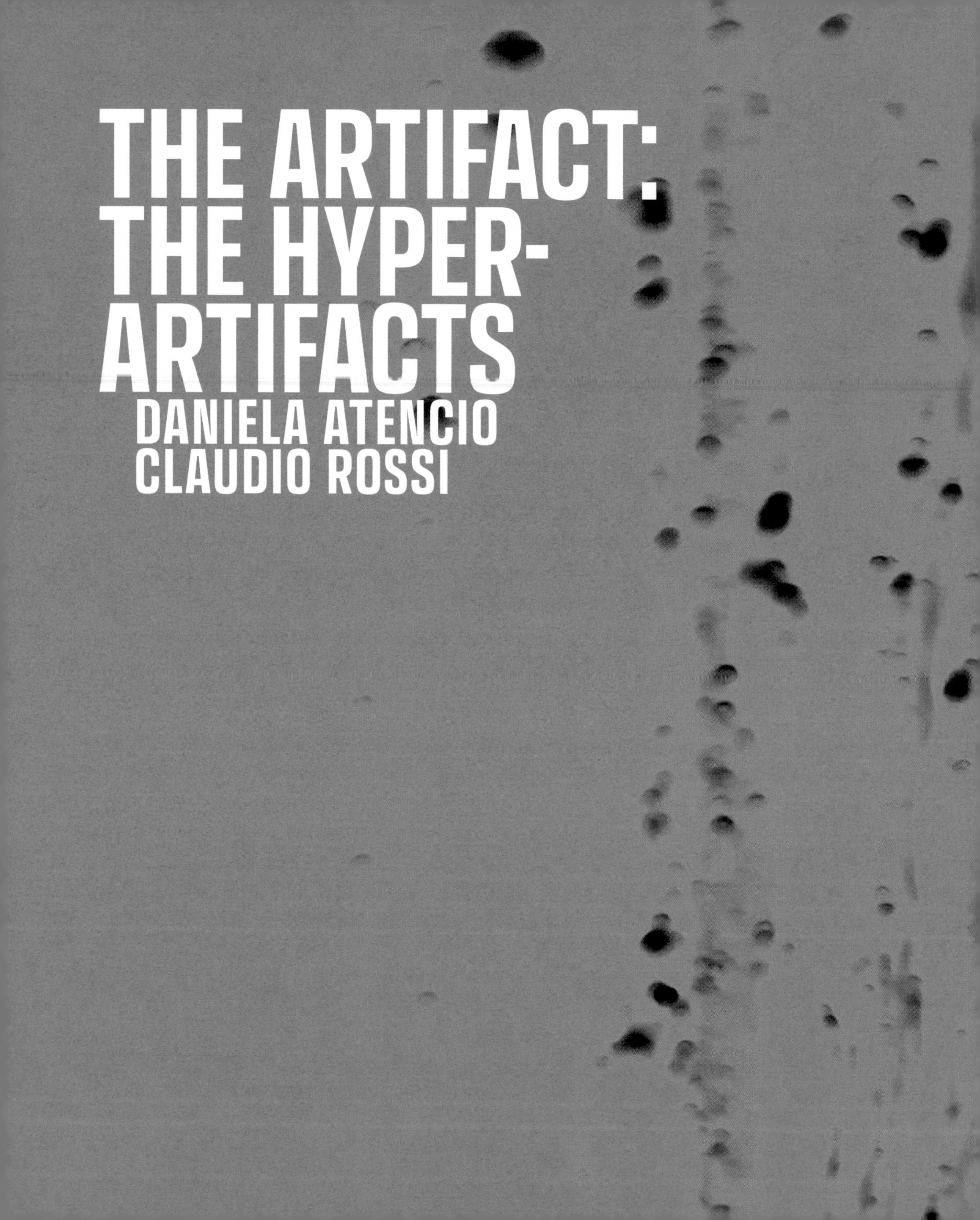

From ~~Artifacts~~ to Hyper-Artifacts

In *The Second Digital Turn*[1], *Mario Carpo* argues that representation and fabrication coexist simultaneously. The author explains that the minimal or discrete details of modernity were born from positions like the cost or the stylistic perpetuity of the ornament[2]. However, the new digital technologies and robotic explorations have given objects and their material expressiveness a turn of relevance—a new way of understanding stylistic and aesthetic ornamentations and organizations, thus raising new questions about materiality and, above all, the new responsibilities, or purposes of the architectural *artifacts*. A new computational discretism and brutalism, terms proposed by *Carpo*[3], seem appropriate to describe this formal expression.

A resolution for the crisis in architectural representation cannot exist without a theory of the artificial and the artifacts. It is necessary to propose a novel toolbox in which the virtual, the real, the physical, and the digital coexist. This process implies inventing and rupturing possible connections, therefore, to challenge these categories through the design exercise, aiming for new definitions for the architectural *artifact*.

In *Robot House*[4], *Peter Testa* proposes a change in the design approach, where protocols and images merge into a contemporary discourse about objects. Projections and mappings collide with different dimensions of reality, constituting what the author calls a new object—in our case, a new *artifact*. The digital and physical projections, matter arrangements, and geometries converge in a new conceptual frame: Objects produce hyper-objects, and *artifacts* produce *hyper-artifacts.*

To reconstruct the definition of the term *artifact*, we must first focus on its etymological root and two of its definitions. The first of these definitions is that the word *artifact* comes from the Latin *arte factum,* or 'made with art'[5]. From this base, we can establish two considerations: first, the relationship with art —which will be developed further insofar in other parts of this text, as this definition responds to philosophical positions; and second consideration is that the word 'made' concerns to the action of 'doing' (*factum*, verb). The second definition is that the word *artifact* thus relates to the understanding of 'facts' as income parts of an experimental or speculative process with reality, but also as the 'truth', or as a physical, beautiful hidden truth.

1 Carpo, M. (2017) "*The Second Digital Turn: Design Beyond Intelligence*". Cambridge: The MIT Press.

2 Loos, A. ([1910] 2019) "*Ornament and Crime: Thoughts on Design and Materials*", New York: Penguin.

3 Carpo, *Op. Cit.*

4 Testa, P. (2017) "*Robot House: The New Wave in Architecture and Robotics*", London: Thames & Hudson.

5 De Vaan, M. (2008) "*Etymological Dictionary of Latin and the other Italic Languages*", Leiden: Boston.

A Beautiful Objectual Experiment

In books VI to XIX of the *De re ædificatoria* treatise[6], *Leon Battista Alberti* explains that beauty is hidden in the material arrangements that constitute the spatial thickness of architectural objects. Beauty, then, is the result of the materialization of specific cultural, political, and social values. The author describes different characteristics using linguistic, rhetorical, and artistic narratives. The material substance is, borrowing his metaphor, "*the meat that covers the bones*"[7]. The spatial thickness determined by its materiality and capabilities refers to that artificial 'meat' that transposes natural conditions and mediates inhabitable situations. It serves as the foundation for experimentation to understand how *artifacts* can have new spatial responsibilities and properties.

Going back to the relationship with art, *Ferrater Mora* establishes in his *Dictionary of Philosophy*[8] that, for the classical Greek authors, one of the conditions for art to be is its ability to imitate nature. As the author states, this "*meant that all authors agreed on the Platonic conception of the work of art as one imitation of an imitation*".[9] Art is then an artifice of the artificial. Is the capacity for transformation of reality into a different state, or, as defined by the author: "*a rational artifice*",[10] or in this case, a rational *artifact.*

Understanding some linguistic definitions[11], we can hold together two of the following meanings for an *artifact*:

m. Object, especially a *machine* or an *apparatus*, built with a certain technique for a certain purpose. (i.e.: An electronic device or a flying device).

4. *m.* In a study or in an *experiment*, a factor that disturbs the correct interpretation of the result.

These definitions correlate with some of the properties expressed by *Milton Nahm* about works of art, translated as the characteristics in the conformation of *artifacts*. For the author, an *artifact* is, from its morphology, a concrete and significant form:

6 Alberti, L.B. (1485) "*De re ædificatoria: Vols. VI–XIX*". Translated by Rykwert, J., Tavernor, R. & Leach, N. (1988), Cambridge: MIT Press.

7 *ibidem*

8 Ferrater Mora, J. (1994) "*Diccionario de Filosofía*", Madrid: Editorial Ariel.

9 *ibidem*

10 *ibidem*

11 Real Academia Española: *Diccionario de la Lengua Española*, 23.ª ed., [versión 23.6 en línea]. <https://dle.rae.es> [2023/07/23]

concrete in terms of its materiality and significant because of its expressiveness[12]. The *artifact* is also an invention: it must be produced through a process of work, study, and experimentation. Based on these considerations and realigning what was proposed by *Testa*[13], the definition of *hyper-artifact*[14] should be set as a hybrid *artifact* or a rational artifice with complex forms, materials, and representations as the results of a speculative design process.

To achieve the development of all these material capacities and arguments, we have elaborated a series of methodological phases that created, step by step, a workflow of pedagogical intentions in the teaching of the architectural project, creating loops that allowed to feed back the design process. We did not intend that this exercise showed a unique and unidirectional work path. We consider that this work is a process for new design protocols and apprehensions based on the use of the robotic arm and technology, as well as a series of specific positions about representation and computational thinking, but above all, a workflow that bets on the construction of knowledge for the future of the Latin-American region.

This was (and still is) the starting point of a beautiful objectual experiment: the ***hyper-artifacts***.

12 Nahm, M. (1956) "*The Artist as Creator: An Essay of Human Freedom*", Baltimore: Johns Hopkins Press.

13 Testa, P. *Op. Cit.*

14 Atencio, D., & Rossi, C. (2023) "*Robot-Lab: Hyper-Artifacts*". Catalog exhibition: Homo Faber 3.0 Appropriations of Digital Fabrication from Latin America, Lima: Editorial UPC.

TRANSLATIONS WORKFLOW: THE ANATOMY OF A HYPER-ARTIFACT

DANIELA ATENCIO

This chapter develops the findings and methodological work process of a series of explorations and approaches through the robotic arm, hybrid representations, and material investigations in the proposition of what have been defined as *hyper-artifacts.* Expressions of the passage of time and material vestiges were investigated through a process by which new architectural objects can be deduced, thus questioning disciplinary positions, and apprehending different material conceptualizations. Deploying a method of objectual manipulation and information gathering through working with different computational and fabrication processes using robotics enabled an understanding of different material capacities throughout different *translations*. This design process provided examples of how history can be better understood by building new objects in a journey that blended the physical and the digital, the past and the future, as well as the agencies involved in these explorations. This research-creation built, through different experimental *translations*, a workflow to define the anatomy of the *hyper-artifact*. The studied cases, as briefly introduced before, were the Cartagena de Indias military wall in Colombia, a beautiful radical ancient military infrastructure that triggered this pedagogical and didactical journey on *Robot-Lab*[1] two years ago; and the Guajira's Desert in the northern region of Colombia, on the frontier with Venezuela, as a landscape platform to tested different scales, materialities, and temporalities.

This chapter also focuses on two issues related to the technological approach and its relationship with design investigations in architecture. First, the research presented here developed specific technological tools and mechanisms—software and hardware[2], to achieve different interactions with a robotic arm, and so deepened scholarly exploration of design strategies that can amplify creative pedagogies for undergraduate architecture students in Latin America. Second, the advanced prototyping in a research-creation process promoted as a disciplinary debate through different speculative procedures to propose a re-conceptualization of the architectural object into *hyper-artifacts*. The Research Through Design methodological objectives became pedagogical objects[3], examining a disciplinary reinterpretation, or reintegration, with both: digital and physical worlds; It also opened contemplation on how learning from conceptual issues and

1 Robot-Lab is described as part of the footnotes in the *Robotic Translations* presented cases of this book.

2 Clear, N. (2014) "*Convergence: Architecture as Integrated Spatial Design*". Cit. In: Spiller, N., Clear, N., "*Educating Architects: How Tomorrow's Practitioners Will Learn Today*". Thames & Hudson (London).

3 Atencio, D., & Rossi, C. (2023) "*Robot-Lab: Hyper-Artifacts*". Catalog Exhibition: Homo Faber 3.0 Appropriations of Digital Fabrication from Latin America, Lima: Editorial UPC.

specific technological technologies can generate new perspectives and promote new inflections and representations.

The debate was, at its most ambitious base, about the past and future of architecture and its encounter with technology in the Latin American region, addressing with a sense of urgency the actual local conditions where it operates. This work constituted a story of the relationships between time and technology, between objects and *artifacts*, or, in terms of this research-creation process, *hyper-artifact's* material capacities, establishing new qualities, arrangements, and, above all, new responsibilities and interpretations as architectural design strategies.

The *hyper-artifacts* probed the tension between assemblage and monolithic readings through design strategies like *Robotic Translations*[4]. This process defied authorship by translating a digitally controlled paths into less predictable materials, producing more tension between the natural phenomena of ruination and an artificial predesigned process. This 'synthetic erosion' procedure edited the seams that articulate the assembly of material parts of an architectural object, unveiling a non-monolithic reading of its anatomy.[5]

Working with various technological agencies implies, from an epistemological standpoint, understanding some disciplinary definitions to be able to produce new, informed hybrid results. As *Mario Carpo* suggests, "*the tools we use inevitably feed back into the things we make*"[6]. As already mentioned, the *hyper-artifacts* that derive from this Research Through Design workflow focused on physical and digital processes, iterations, and prototypes. These processes are adopted as translations in the workflow. An act of research-creation that then presents concepts and processes related to computational design from the *translation* between images and *hyper-artifacts and the* and the manipulation and representation of these operations using the robot.

It comprised a series of operations and performativities that, beyond exploring novel forms and their readings, critically examine how digital and robotic tools can transcend visualization and advance architectural discourse through design processes. In this sense, the workflow was conceptually structured in five *translations*: analogue, digital, robotic (physical), inverse, and the translation assemblies (or the assembly of the *hyper-artifact*)

Returning as a reference to what was conceptualized in the previous section, we speak of translations and mechanisms in the development of the workflow precisely because of the sequentiality of each system and parts developed on each of the translations. More precisely, beyond its own conceptual capabilities, the order, and the content of each mechanism, activated every subsequent step to reach, as a result, the hybrid informed apparatus or *hyper-artifact*.

The students worked in the *Robot-Lab* academic space following the sequentiality of the workflow without determining the final result, which was the ultimate goal of the design exercise. On the contrary, they understood the *translations* as those mechanisms that could inform the idea of the project, and better yet, they went through each part of the process as a journey towards the material, spatial, and formal capabilities of their proposals, summoning new responsibilities and reflections. Thus, the result was unexpectedly realistic. It carried out in its anatomical body, parts, systems, thicknesses, and complexity. It was surrounded by the 'flesh' that *Alberti* mentioned in his treatises.

A consistent condition of the process of working with the workflow was precisely to attend to the specific space of each translation as partial results of a journey that, in the medium term, added up all the reflections of each stage. Also, with the purpose of placing the robotic arm as the main agent of this research-creation exercise, but on the other hand, to give it agency from (and for) the design process itself.

4 See *Robotic Translations: Ruins, and Landscapes*.

5 *ibidem*

6 Carpo, M. (2014) "*Breaking the Curve: Big Data and Design*". Artforum, 169–173.

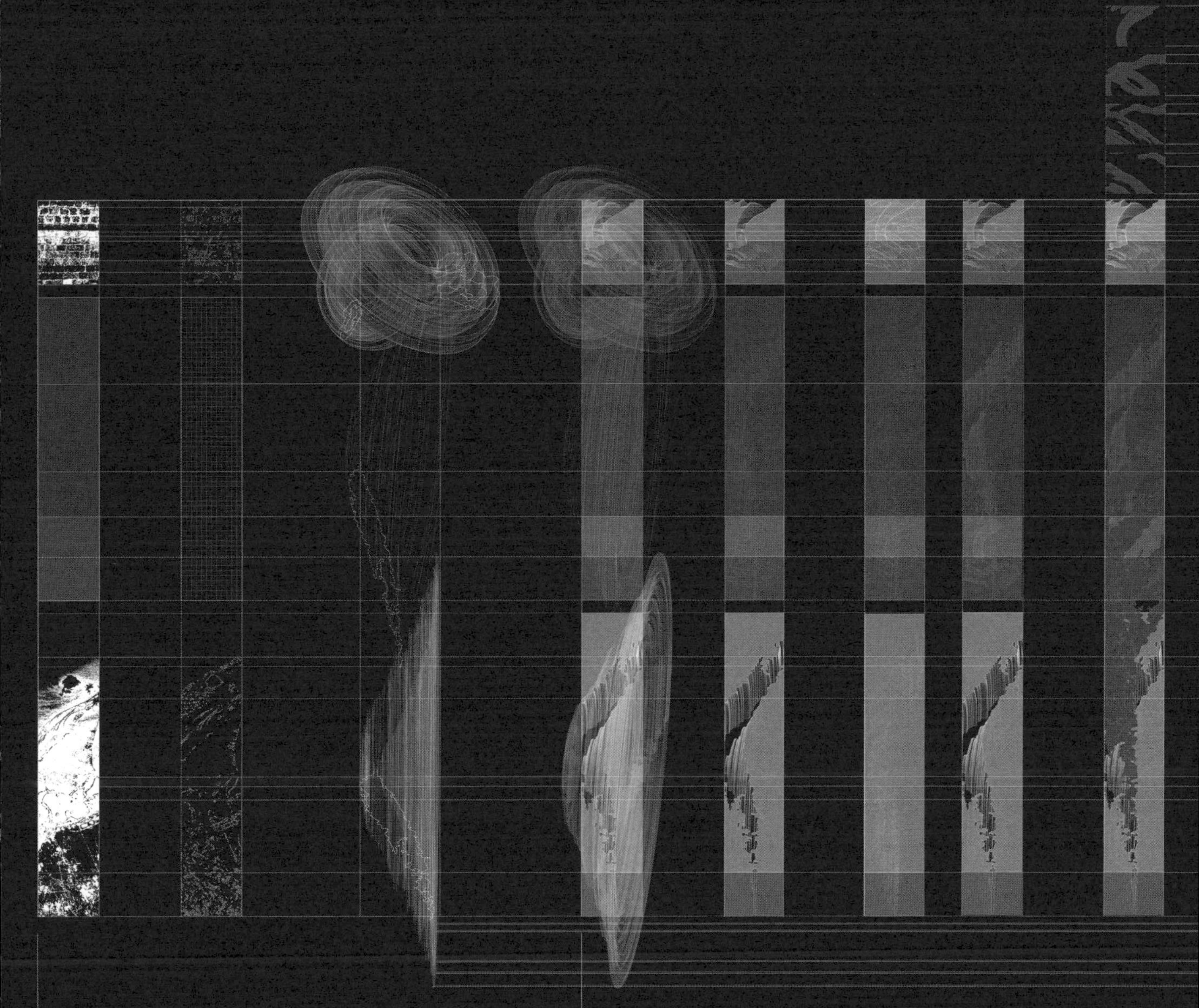

PRESENT AND FOLLOWING TWO PAGES:
TRANSLATIONS WORKFLOW DIAGRAM IN WORM-EYE-OBLIQUE VIEW. ABOVE: RUINS-CARTAGENA DE INDIAS MILITARY WALL APPLIED WORKFLOW. CENTER: OVERALL TRANSLATIONS. BELOW: LANDSCAPES-LA GUAJIRA DESERT APPLIED WORKFLOW.

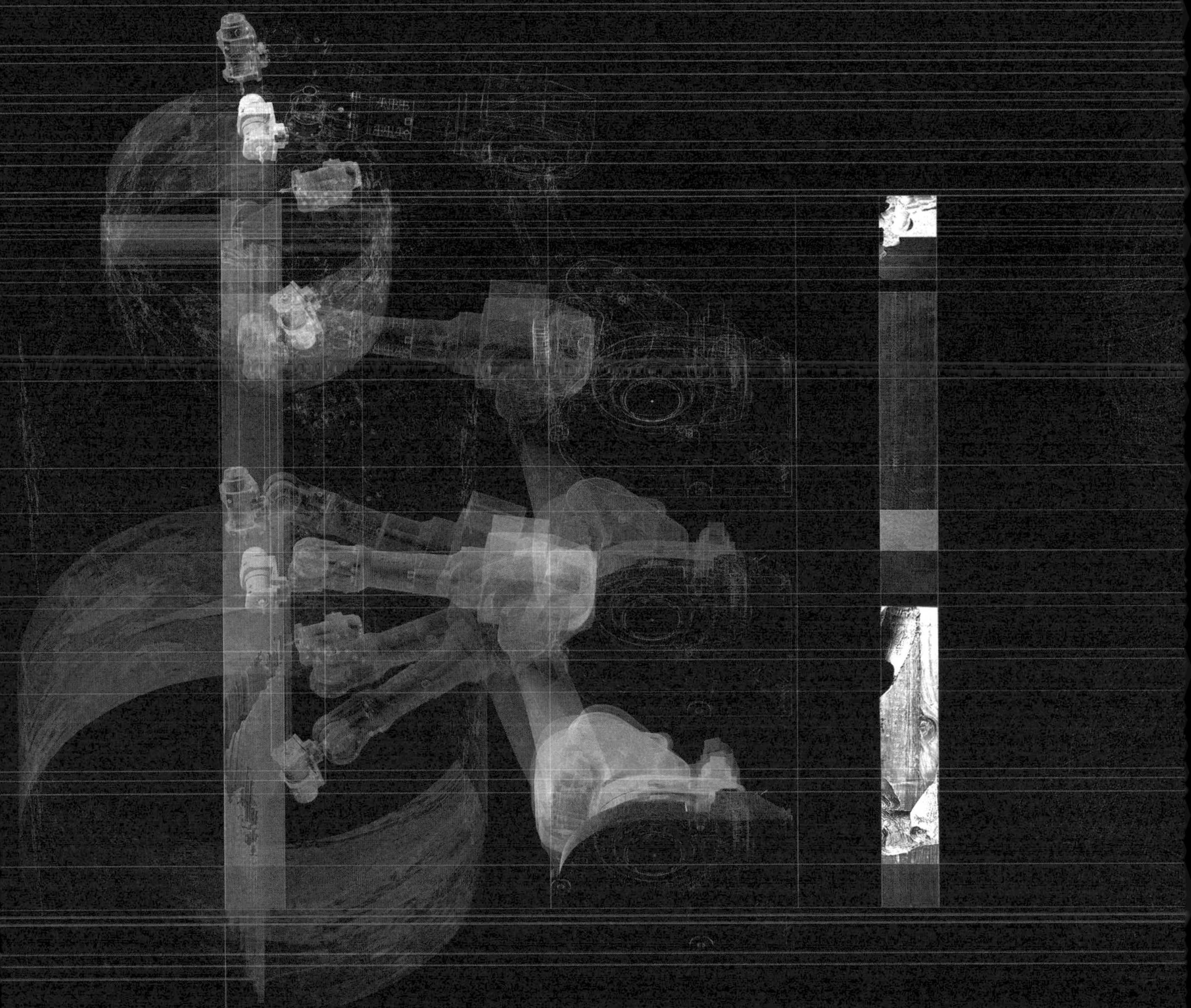

TRANSLATIONS ASSEMBLY — RECONSTRUCTION

MATERIAL SUPERPOSITION

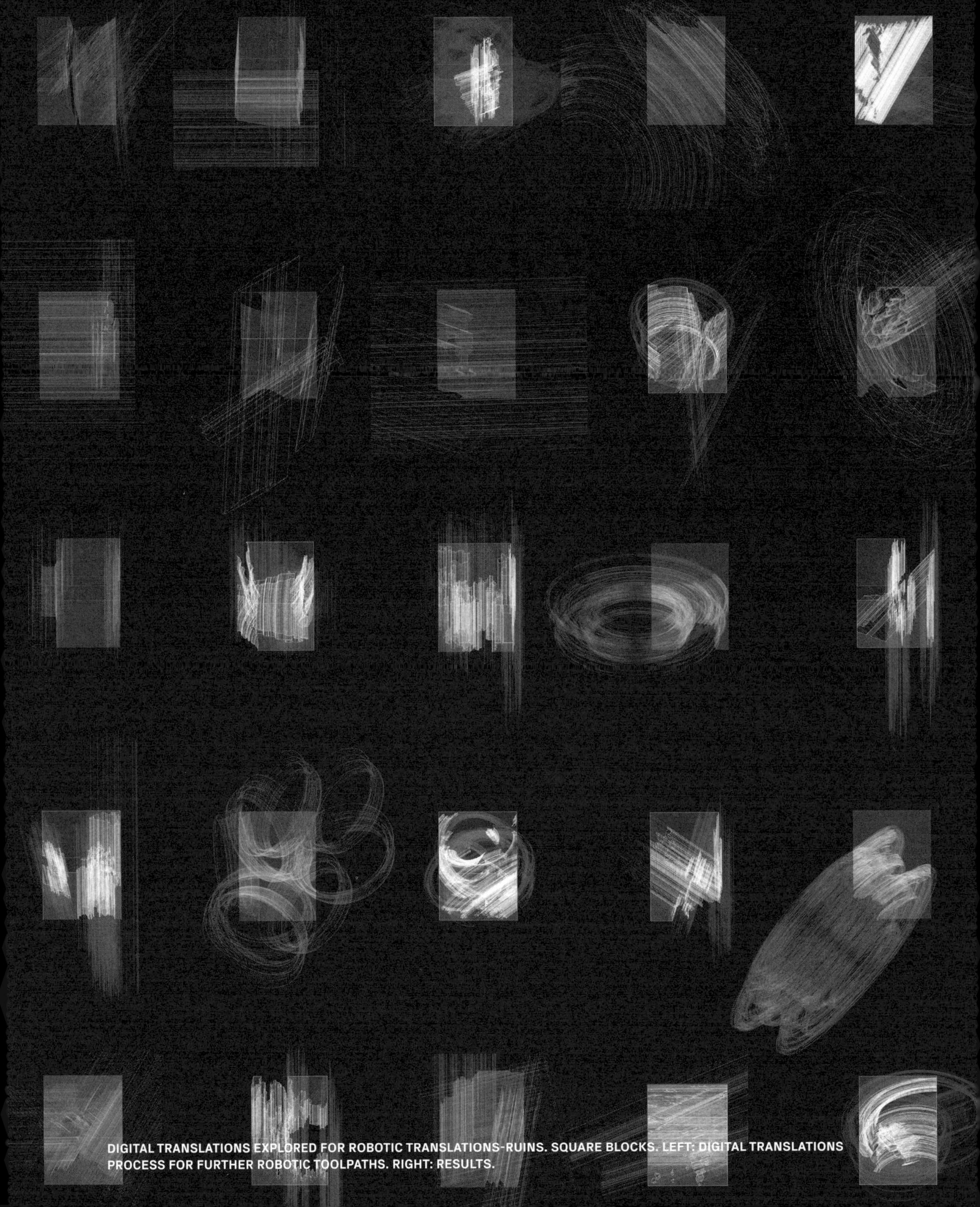

DIGITAL TRANSLATIONS EXPLORED FOR ROBOTIC TRANSLATIONS-RUINS. SQUARE BLOCKS. LEFT: DIGITAL TRANSLATIONS PROCESS FOR FURTHER ROBOTIC TOOLPATHS. RIGHT: RESULTS.

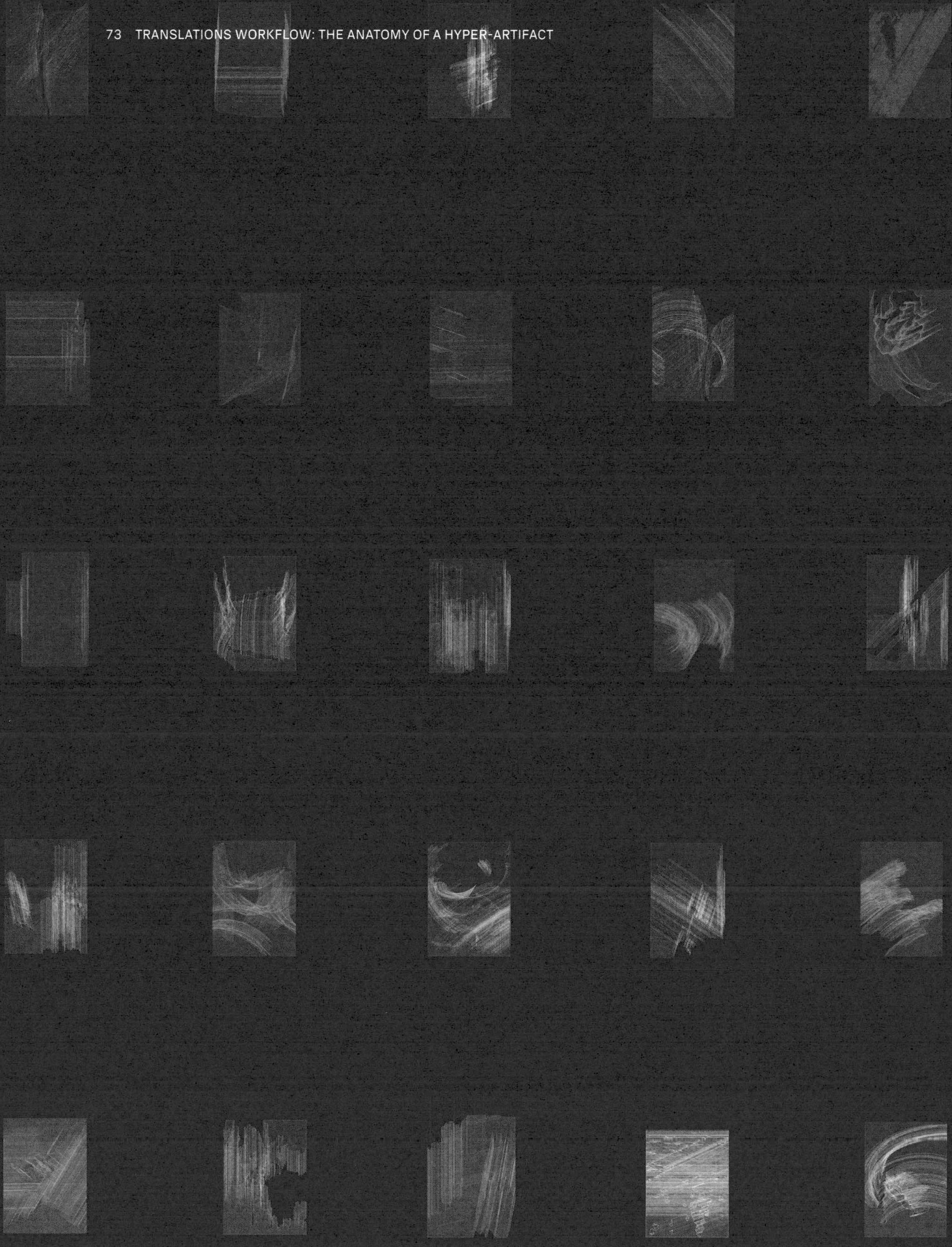

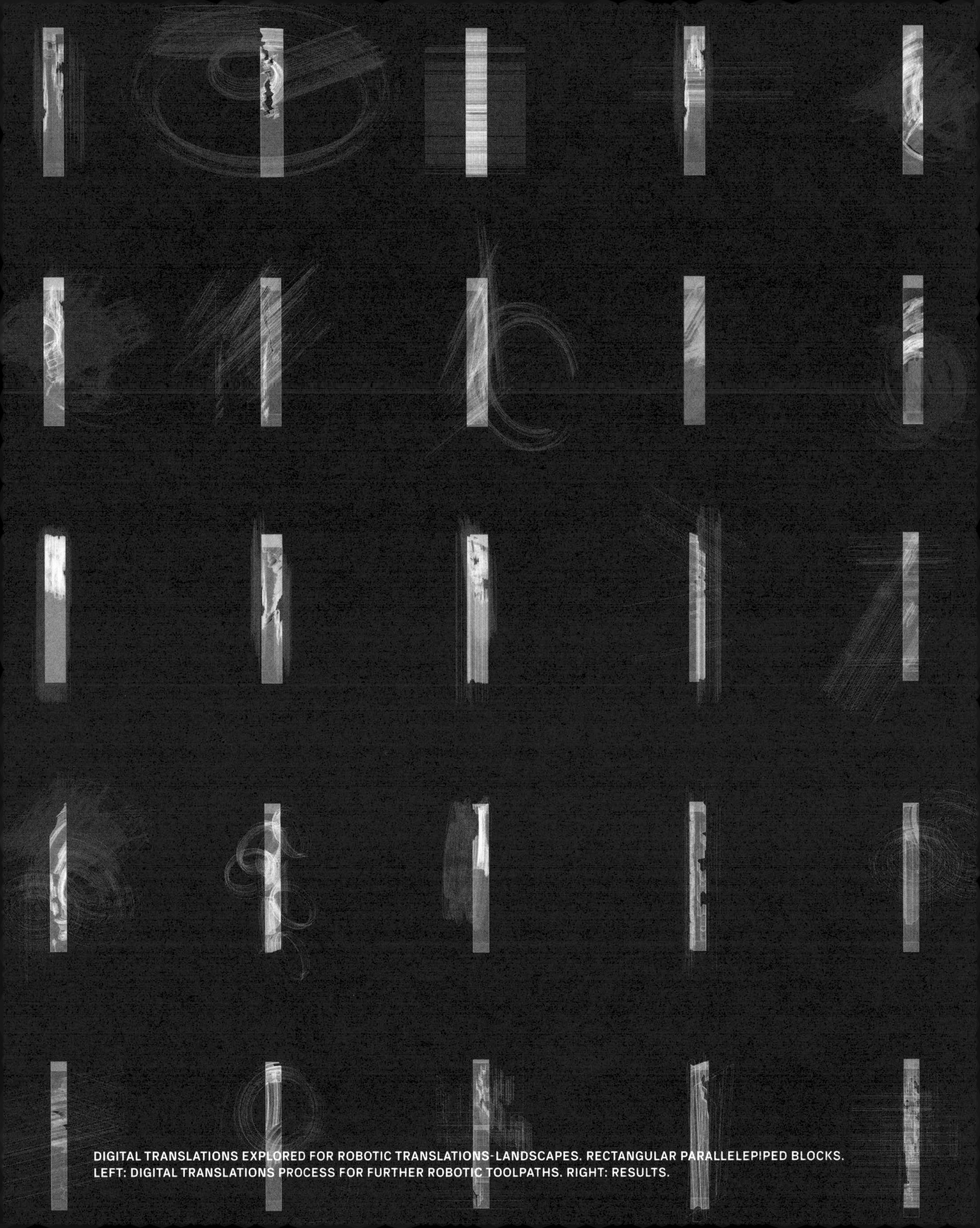

DIGITAL TRANSLATIONS EXPLORED FOR ROBOTIC TRANSLATIONS-LANDSCAPES. RECTANGULAR PARALLELEPIPED BLOCKS.
LEFT: DIGITAL TRANSLATIONS PROCESS FOR FURTHER ROBOTIC TOOLPATHS. RIGHT: RESULTS.

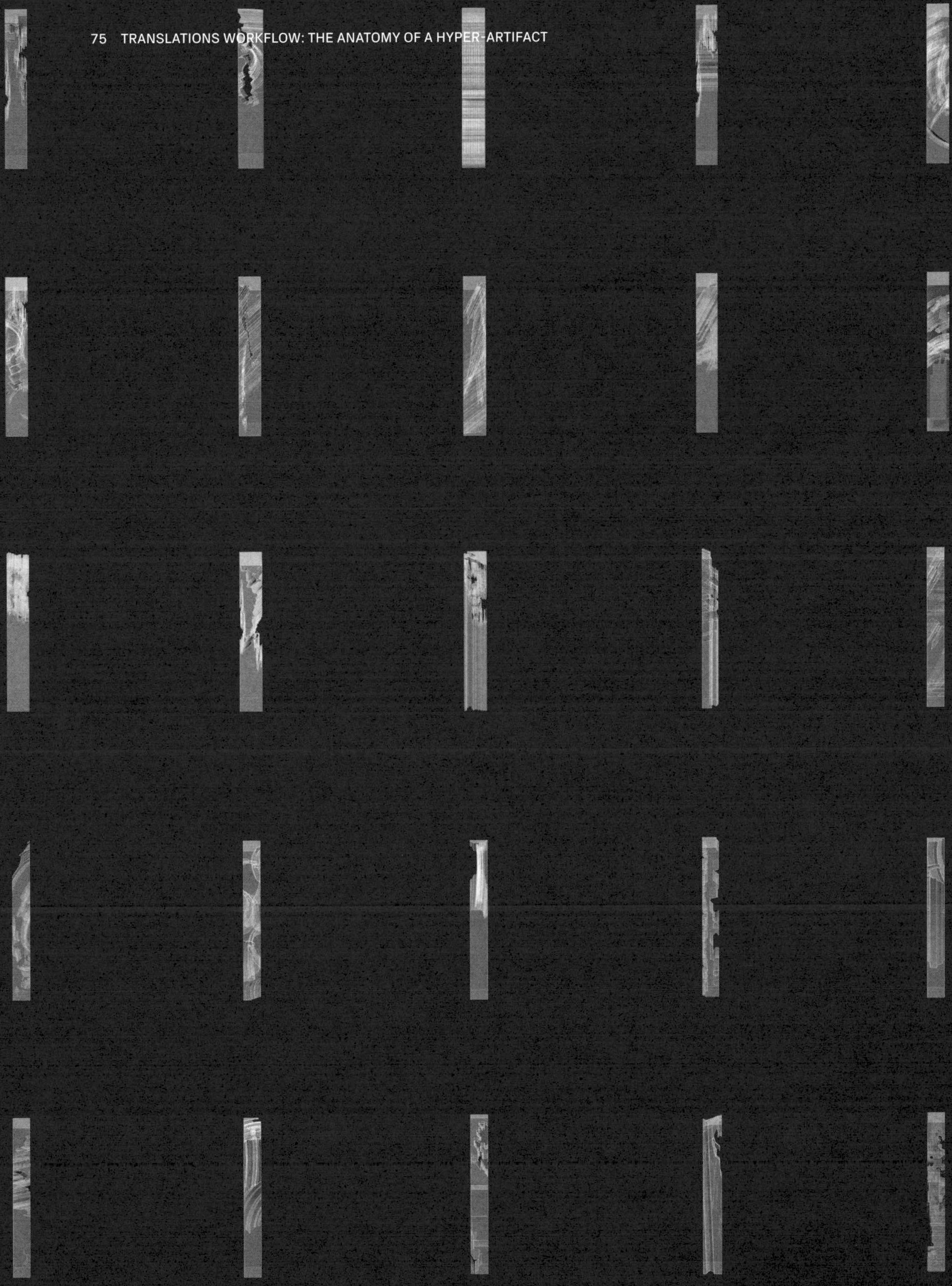

ROBOTIC [PHYSICAL] TRANSLATIONS EXPLORED ON SQUARE BLOCKS [STILL]. ABB 1600. TOOL: MOTOTOOL. EXCERPT FROM ROBOT_LAB: HYPER_ARTIFACTS. HOMO FABER 3.0 . HOMO FABER 3.0 EXHIBITION_APPROPRIATIONS OF DIGITAL FABRICATION FROM LATIN AMERICA_ XXVI SIGRADI INTERNATIONAL CONFERENCE OF THE IBERO-AMERICAN SOCIETY OF DIGITAL GRAPHICS: CRITICAL APPROPRIATIONS.

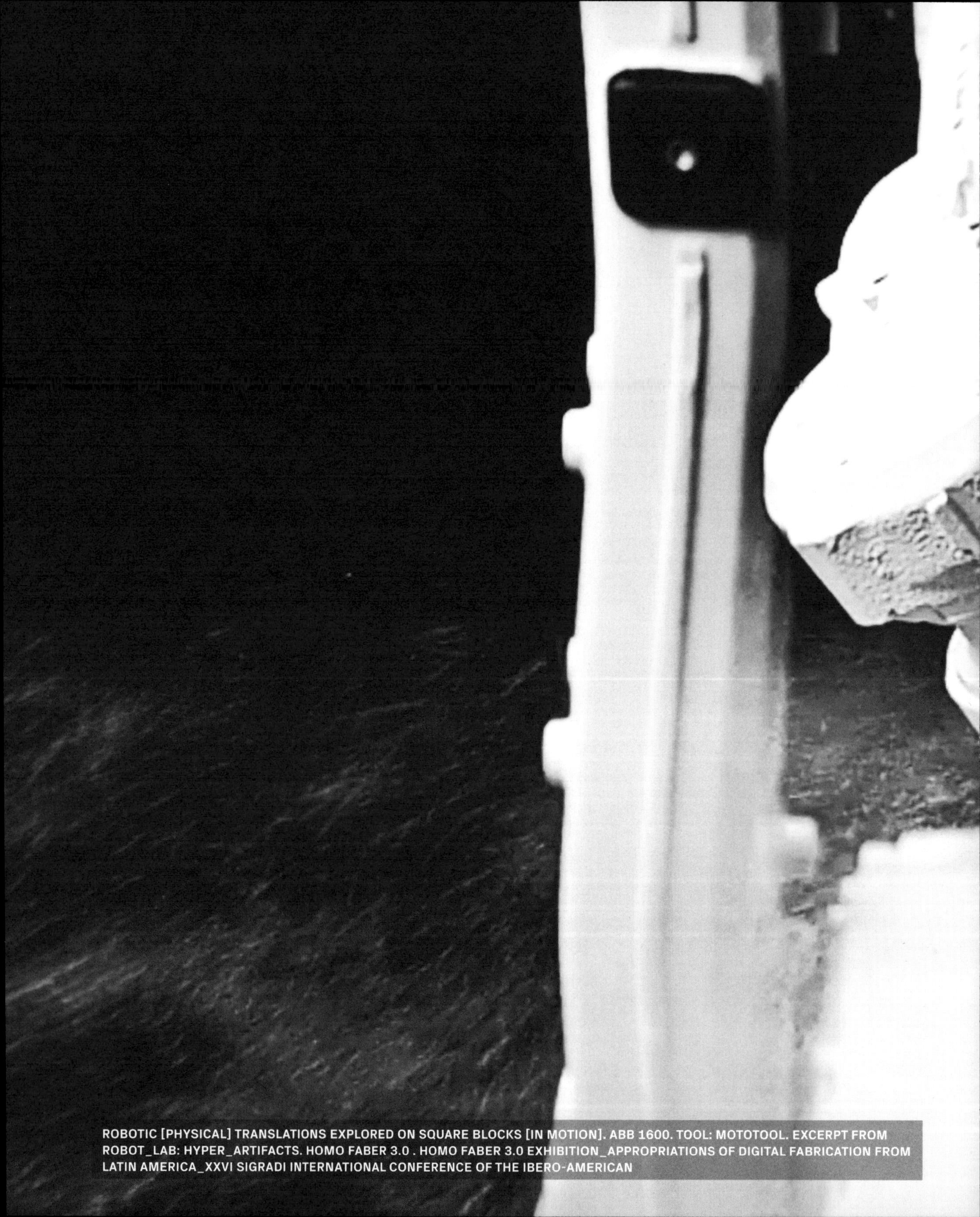

ROBOTIC [PHYSICAL] TRANSLATIONS EXPLORED ON SQUARE BLOCKS [IN MOTION]. ABB 1600. TOOL: MOTOTOOL. EXCERPT FROM ROBOT_LAB: HYPER_ARTIFACTS. HOMO FABER 3.0 . HOMO FABER 3.0 EXHIBITION_APPROPRIATIONS OF DIGITAL FABRICATION FROM LATIN AMERICA_XXVI SIGRADI INTERNATIONAL CONFERENCE OF THE IBERO-AMERICAN

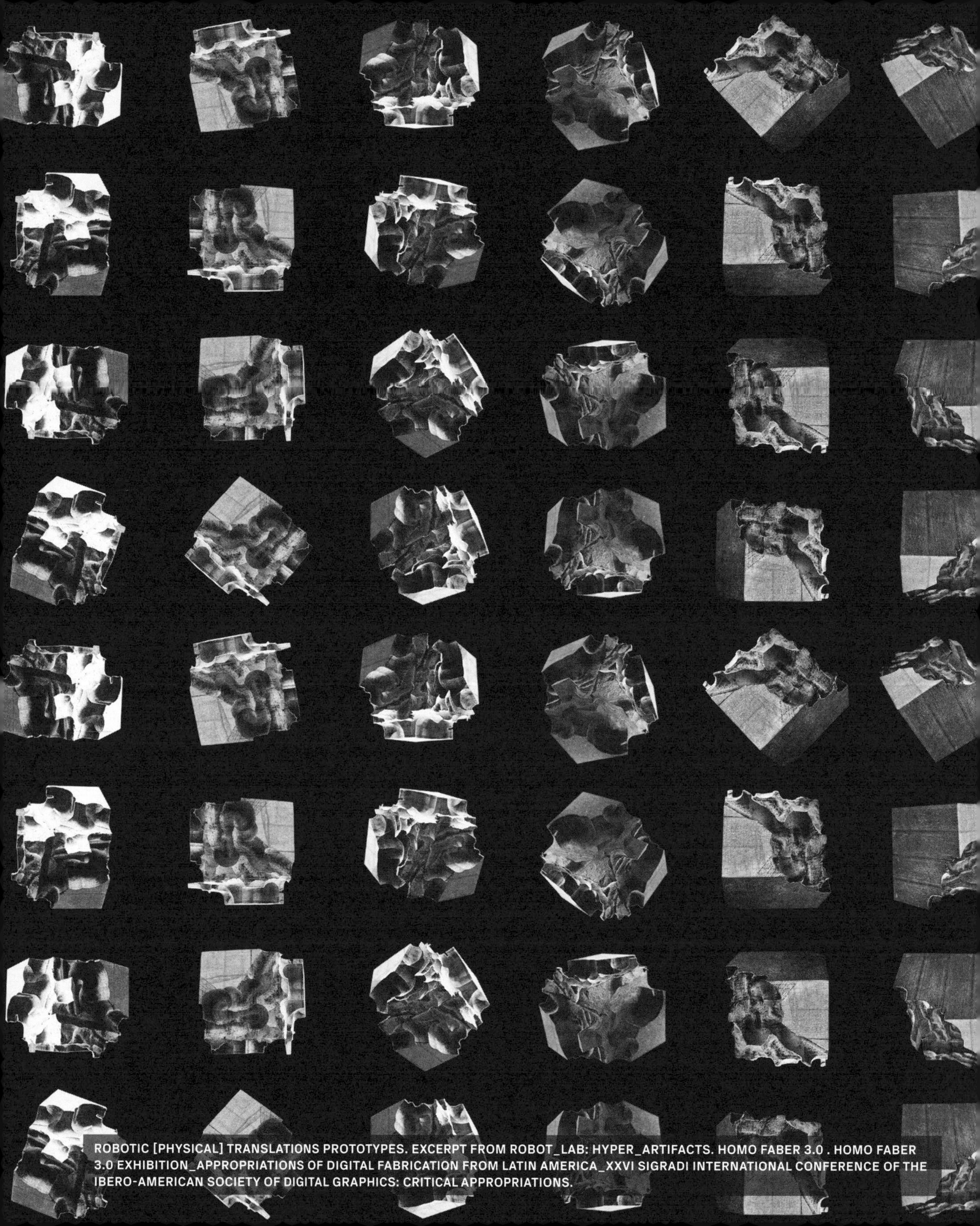

ROBOTIC [PHYSICAL] TRANSLATIONS PROTOTYPES. EXCERPT FROM ROBOT_LAB: HYPER_ARTIFACTS. HOMO FABER 3.0 . HOMO FABER 3.0 EXHIBITION_APPROPRIATIONS OF DIGITAL FABRICATION FROM LATIN AMERICA_XXVI SIGRADI INTERNATIONAL CONFERENCE OF THE IBERO-AMERICAN SOCIETY OF DIGITAL GRAPHICS: CRITICAL APPROPRIATIONS.

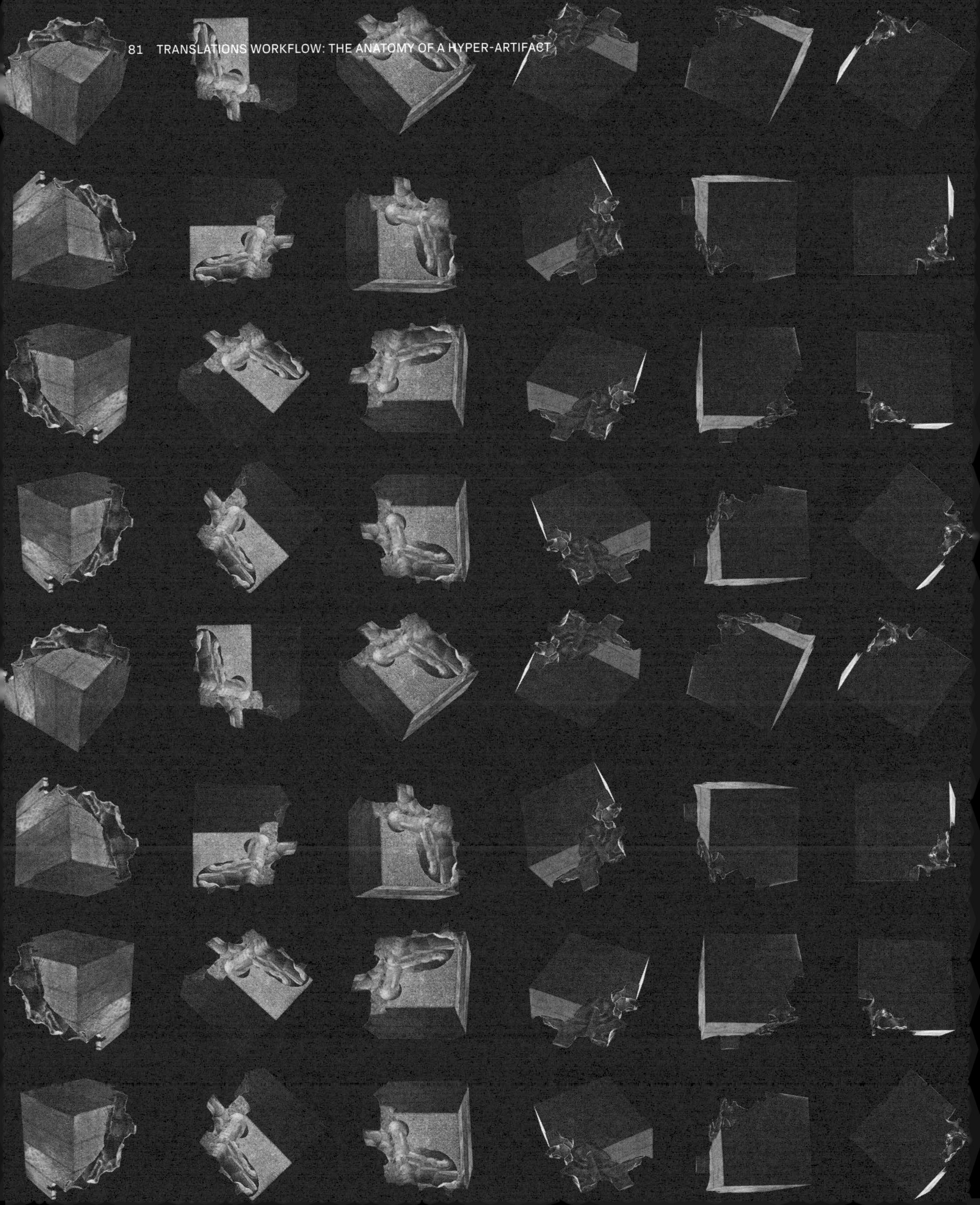

ROBOTIC [PHYSICAL] TRANSLATIONS. EXPLORED ON RECTANGULAR AND SLIM PARALLELEPIPED BLOCKS [VERTICAL]. CORNERS ROBOTIC TOOLPATHS APPLIED. ABB 1600. TOOL: MOTOTOOL.

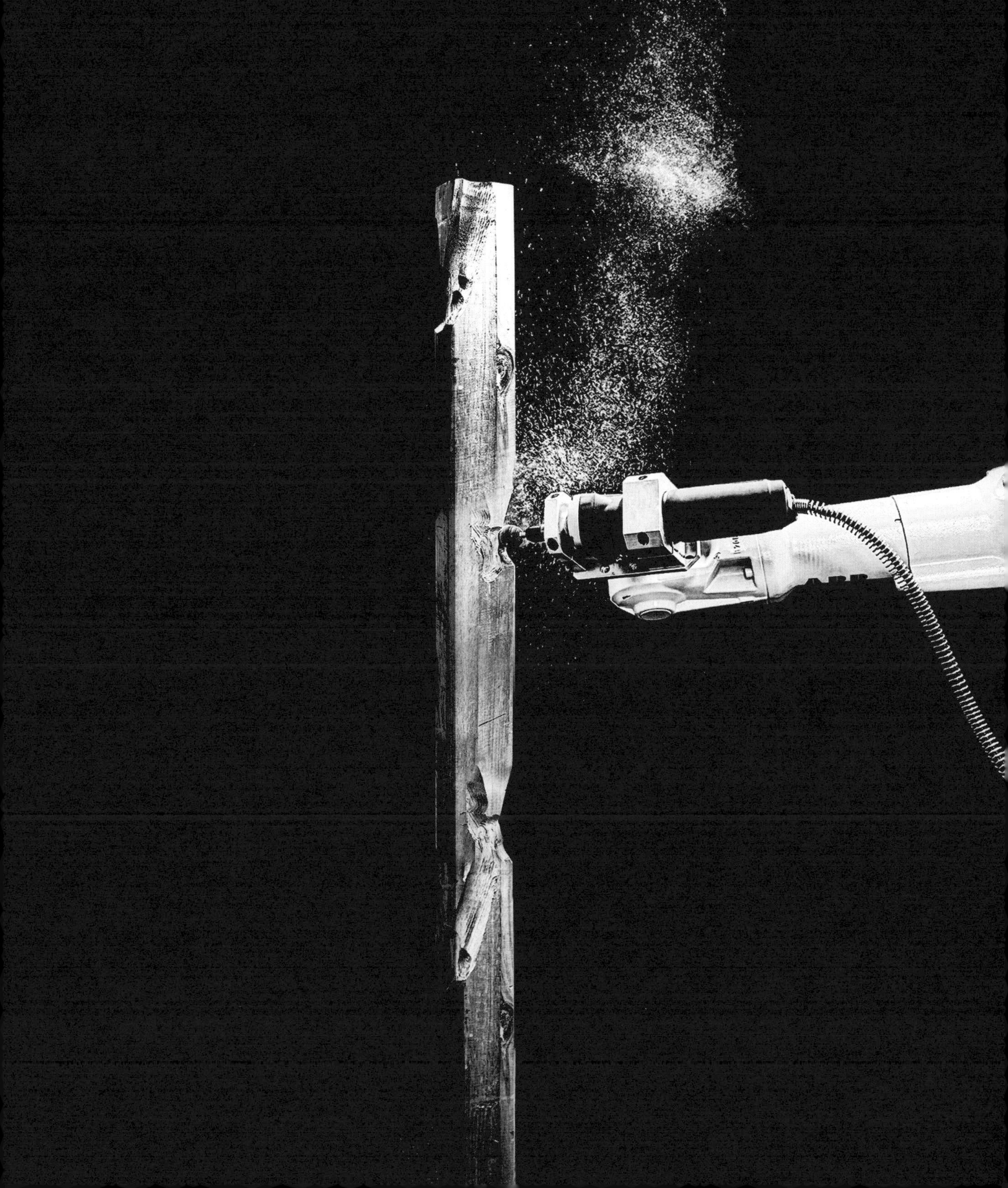

ROBOTIC [PHYSICAL] TRANSLATIONS. EXPLORED ON RECTANGULAR AND SLIM PARALLELEPIPED BLOCKS [HORIZONTAL]. CORNERS ROBOTIC TOOLPATHS APPLIED. ABB 1600. TOOL: MOTOTOOL.

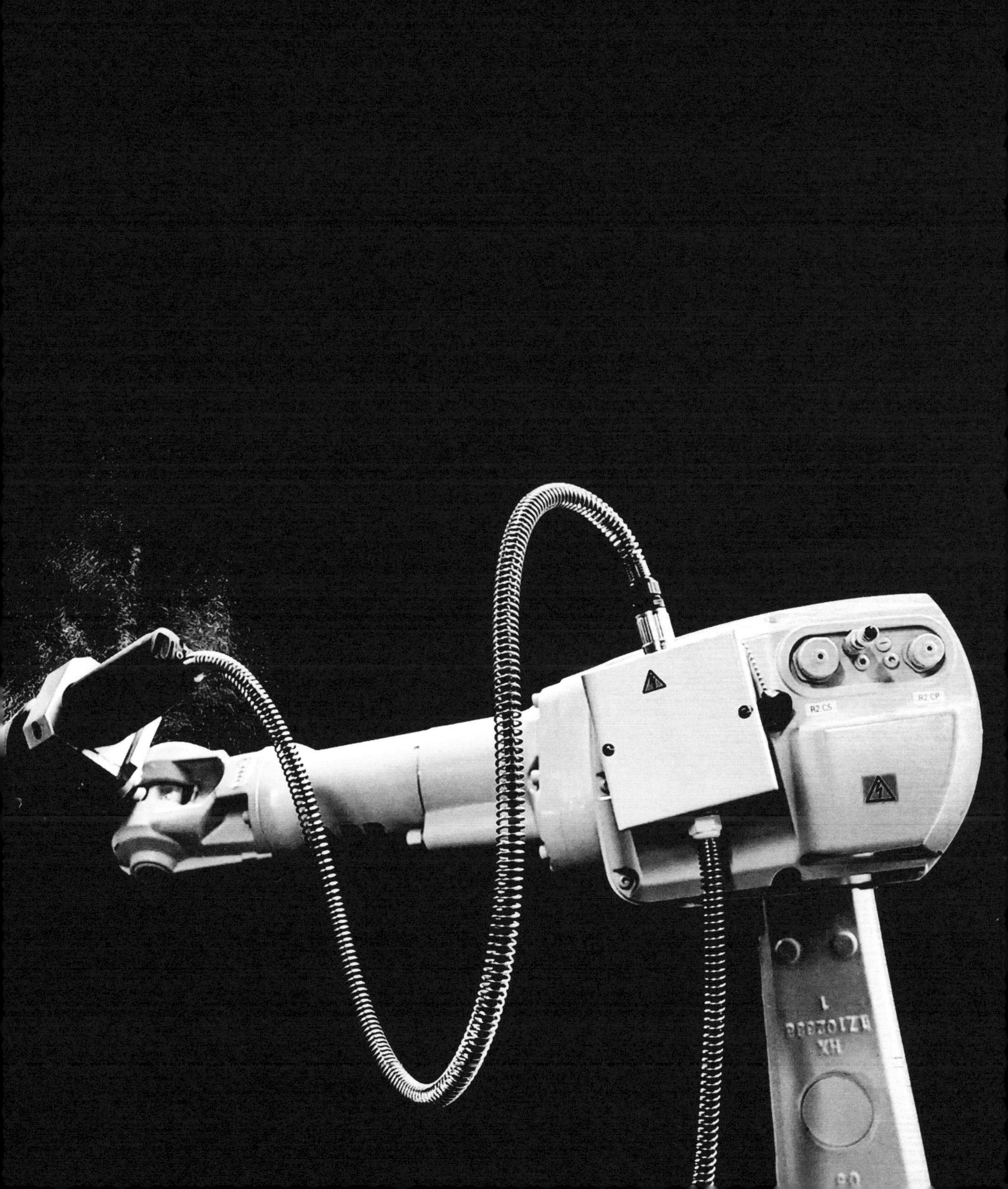

ROBOTIC [PHYSICAL] TRANSLATIONS. EXPLORED ON RECTANGULAR AND SLIM PARALLELEPIPED BLOCKS [SUSPENDED]. CORNERS ROBOTIC TOOLPATHS APPLIED. ABB 1600. TOOL: MOTOTOOL.

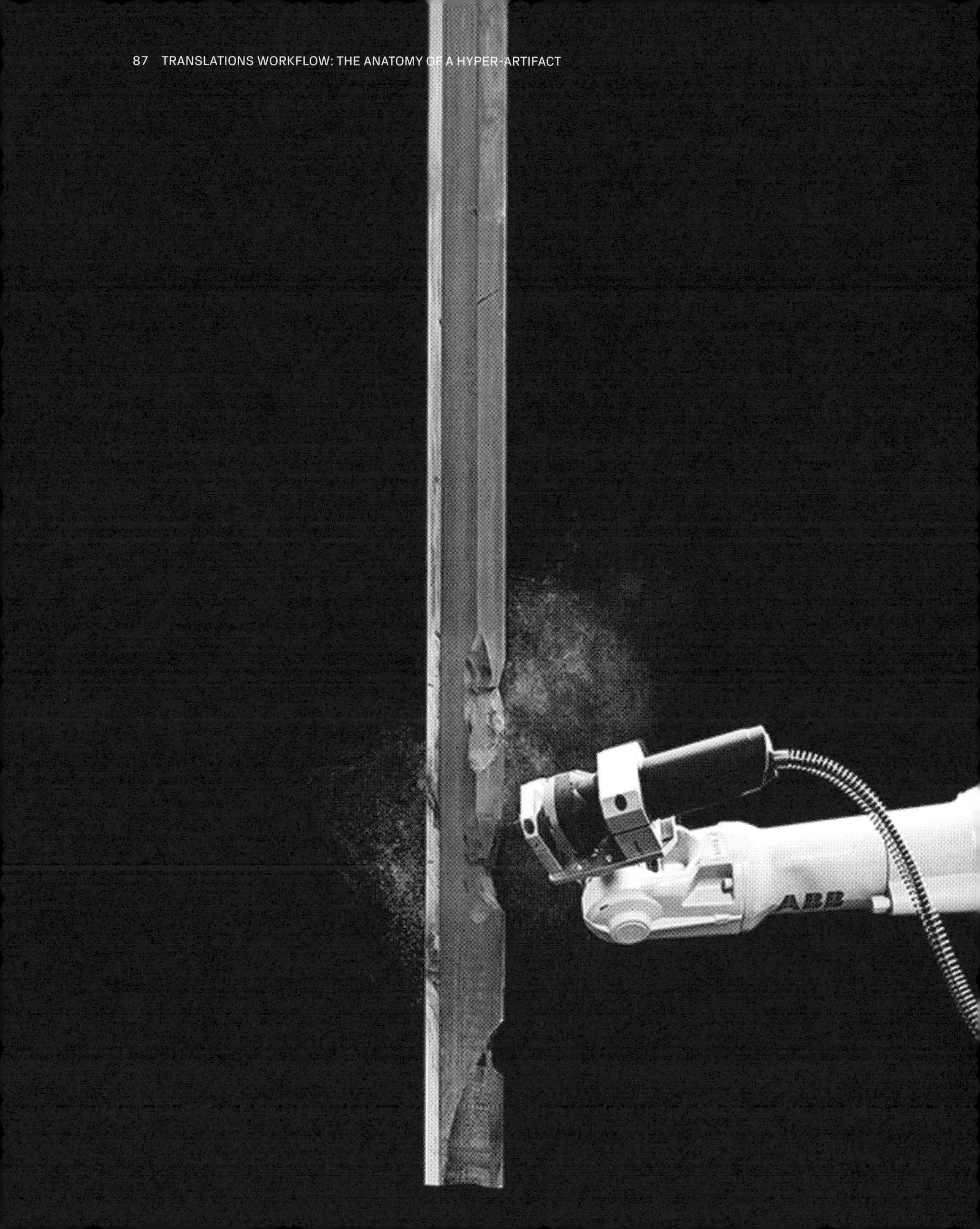
ABB

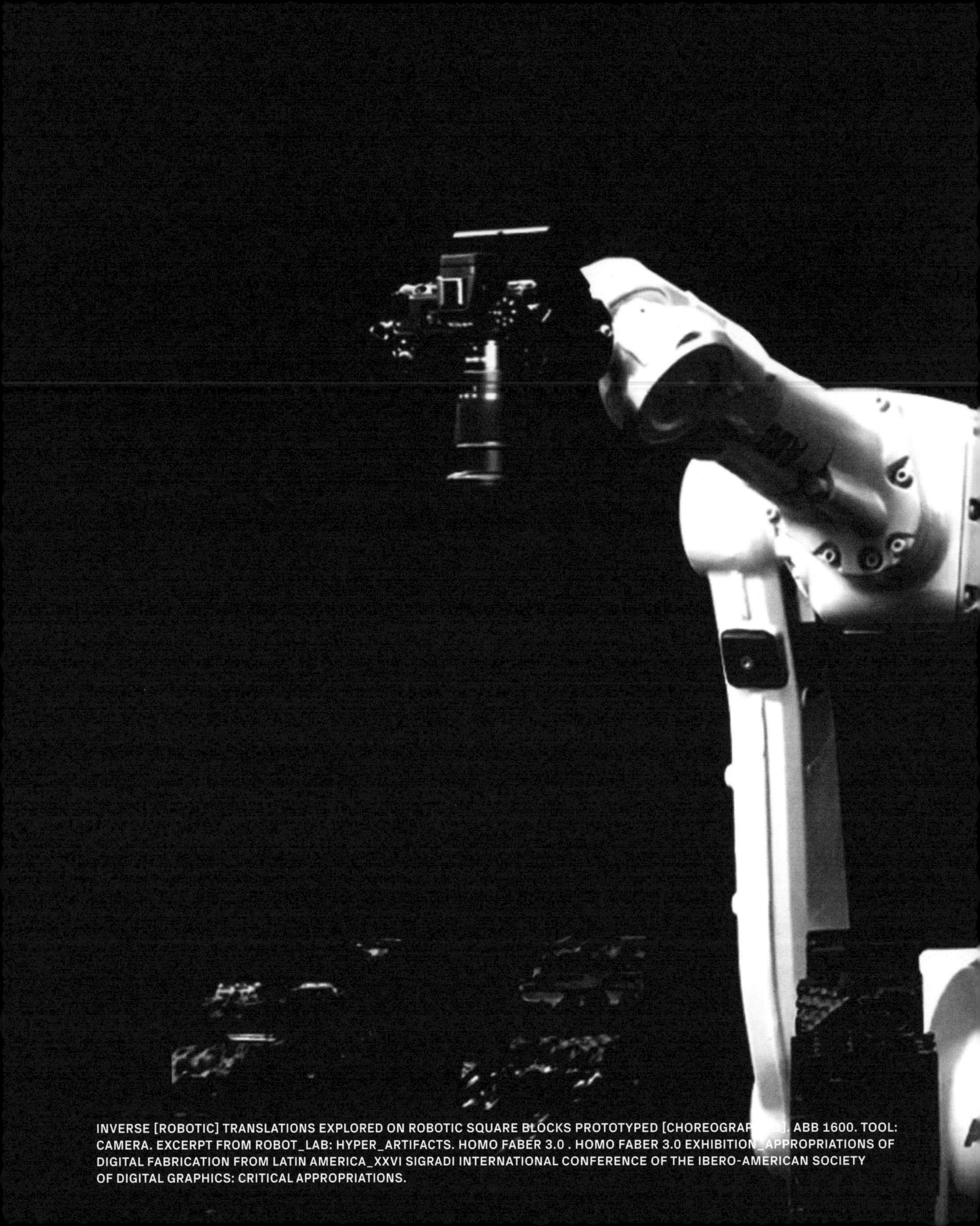

INVERSE [ROBOTIC] TRANSLATIONS EXPLORED ON ROBOTIC SQUARE BLOCKS PROTOTYPED [CHOREOGRAP[illegible]]. ABB 1600. TOOL: CAMERA. EXCERPT FROM ROBOT_LAB: HYPER_ARTIFACTS. HOMO FABER 3.0 . HOMO FABER 3.0 EXHIBITION_APPROPRIATIONS OF DIGITAL FABRICATION FROM LATIN AMERICA_XXVI SIGRADI INTERNATIONAL CONFERENCE OF THE IBERO-AMERICAN SOCIETY OF DIGITAL GRAPHICS: CRITICAL APPROPRIATIONS.

INVERSE [ROBOTIC] TRANSLATIONS EXPLORED ON ROBOTIC SQUARE BLOCKS PROTOTYPED [CHOREOGRAPHY 2]. ABB 1600. TOOL: CAMERA. EXCERPT FROM ROBOT_LAB: HYPER_ARTIFACTS. HOMO FABER 3.0 . HOMO FABER 3.0 EXHIBITION_APPROPRIATIONS OF DIGITAL FABRICATION FROM LATIN AMERICA_XXVI SIGRADI INTERNATIONAL CONFERENCE OF THE IBERO-AMERICAN SOCIETY OF DIGITAL GRAPHICS: CRITICAL APPROPRIATIONS.

INVERSE [ROBOTIC] TRANSLATIONS EXPLORED ON ROBOTIC SQUARE BLOCKS PROTOTYPED [CHOREOGRAPHY 3]. ABB 1600. TOOL: CAMERA. EXCERPT FROM ROBOT_LAB: HYPER_ARTIFACTS. HOMO FABER 3.0 . HOMO FABER 3.0 EXHIBITION_APPROPRIATIONS OF DIGITAL FABRICATION FROM LATIN AMERICA_XXVI SIGRADI INTERNATIONAL CONFERENCE OF THE IBERO-AMERICAN SOCIETY OF DIGITAL GRAPHICS: CRITICAL APPROPRIATIONS.

ANALOGUE TRANSLATIONS

DANIELA ATENCIO

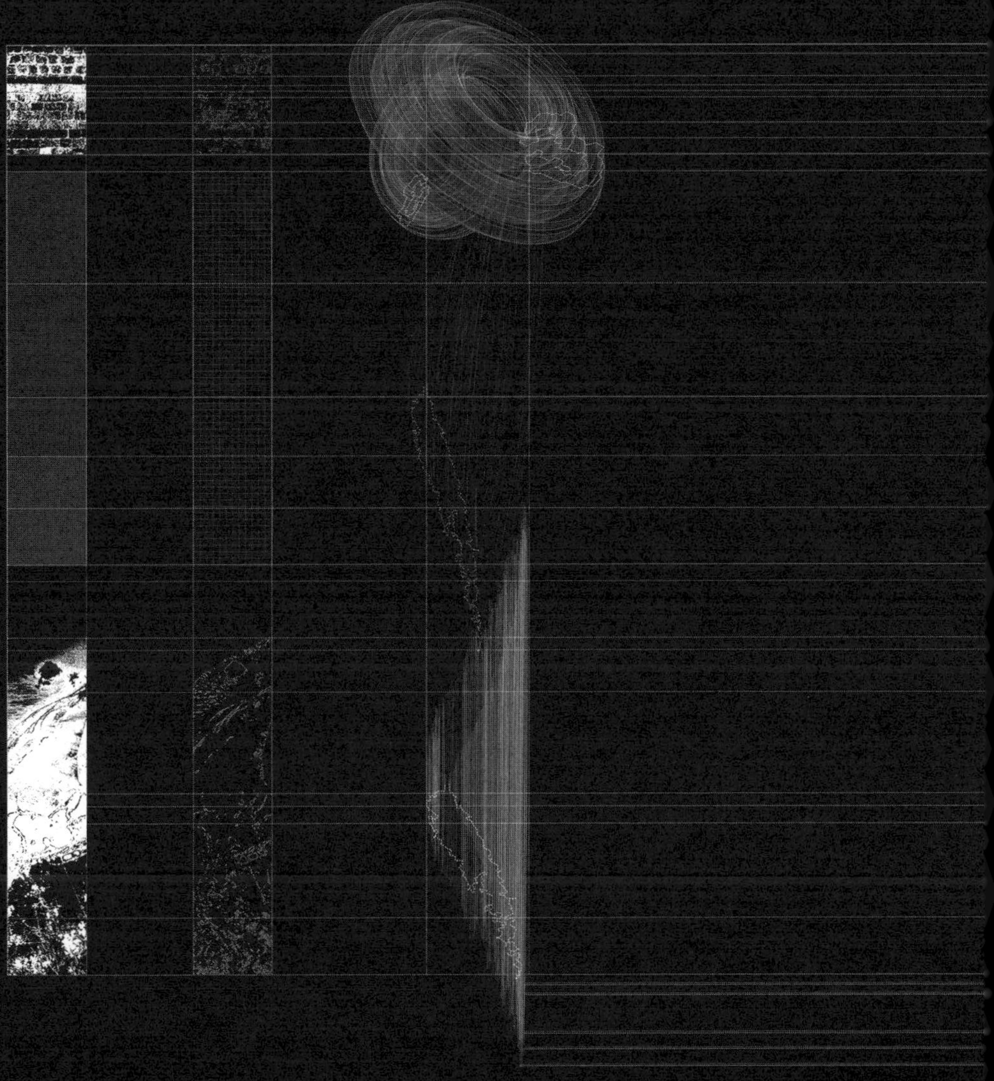

ANALOGUE TRANSLATIONS DIAGRAM IN WORM-EYE-OBLIQUE VIEW. ABOVE: ROBOTIC TRANSLATIONS-RUINS. CENTER: OVERALL TRANSLATIONS. BELOW: ROBOTIC TRANSLATIONS-LANDSCAPES.

Developing each translation and its sequences created a pedagogical-didactic path that challenged the digital physical relationship. Unlike the criticism of the separation of parts, the analogue alludes to a new anatomical conception of the studied objects and approaches to the description of the hidden capacities.[1] In this procedure, material resolutions, agencies, arrangements, elements, and parts are some of the aspects addressed to understand their qualities and capacities, as well as the properties that emerge from their relationships. The trigger for this workflow was selecting the object of study (e.g.: Cartagena's military wall), to do a detailed registration and observation. The representation and its modeling described the specificity to lead to a detailed reading of its constitutive 'anatomical body' of each case studied.

The anatomy of the *Hyper-artifact*[2], mentioned in previous chapters, it's a system of technical elements and constructive pieces informed, as a starting point, from the analogue understanding of the studied case; also, is a compositional conjunction of typological, morphological, and syntactic readings, from symbolic, perceptual, cultural, and/or social-political narratives. It is an *analogue translation* since the accumulation of different processes in its architectural thickness, testifies the deterioration, friction, erosion, or ruination, in other words, the analogue reading of time in a *taxonomic artifact.* This artifact become the exercise of conceptualizations and descriptions of all the parts and the whole, thus patterning the descriptive observations of human anatomy treatises of the 19th century (e.g.: *Testut and Latarjet*[3]).

These components and processes are captured through photography (static image registration) and 3DLiDAR scanning from drone flights and other LiDAR equipment (dynamic image registration). After capturing all this imagery, the next step in the workflow was to 3Dmodel each part and element with meshes, NURBS, and/or SubD's, using a mix of 3Dmodeling and point cloud software processes. With these mechanisms, all the geometries and specificities were digitally captured, so the first translation relates to the journey from real physical data to the digital world through images.

Paradoxically, to rethink the existing artifacts, the focus is on visual thinking and non-linguistic technologies. This argument is consistent with *Peter Sloterdijk's* ideas that today's most advanced thinking must make use of 'imaging' techniques to "*illuminate the landscapes of ideas, discourse, and data [...] with a new kind of conscious formal vision.*"[4] For the author, an image is not a representation of an idea, but

1 Latour, B. (2007) "*Can We Get Our Materialism Back, Please?*", ISIS, 98(1), 138–142.

2 See chapter *Hyper-artifact*

3 Testut, J.L., & Latarjet, A. ([1928] 1982) "*Tratado de Anatomía Humana*". Barcelona: Salvat.

4 Sloterdijk, P. (1998) "*Spheres*". Frankfurt: Suhrkamp Verlag.

makes sense from an "*image signifying ideas*,"[5] leaving aside materialist and phenomenological ontologies. *Henri Bergson* early discovery of matter as a set of images and his conceptual framing of an "*image of movement*"[6] and, more profoundly, of an "*image of time*,"[7] is essential to understand the taxonomy of the cases studied.

Bergson demonstrated the various ways in which the image challenged modernist empiricism and its reliance on abstract geometry to define matter. This *analogue translation*, which refers to physical, synchronous, and diachronic processes, is affected by the relationship between image data and technological interpretations. The modeling of the parts not only referred to the observation of the evident physical reality but also to unveil the hidden dimensions, extracted from the images. A processing of reciprocal information between reality and the connection with the digital environment.

Also, modeling the totality of the existing artifact and its different compositional elements, allowed to build an exploited image to understand their assembly process analogue taxonomy. Each part is analyzed to recognize correlations and non-correlations with the studied object. It was crucial to identify singularities as unique pieces and to study from this unique characterization the artifact's possible, eminent, and emerging properties. In this sense, this procedure recalls the observation of *Bruno Latour* in *Can We Get Our Materialism Back, Please?*[8] about *Damián Ortega's* installation: *Cosmic Thing*.

The transparency offered by the exploited view of the installation, establishes two connective viewpoints to the analogue translation. On the one hand, the specific understanding of that each part creates an inventory of objects that gives the VW vehicle, in Ortega's case, an objectual thickness with functional logics. This, which *Latour* called 'thingness',[9] was one of the operative and conceptual analogies for this workflow. On the other hand, the observation on a totality composed of parts, conformed the other material capacity: 'thicknesses'.

The outcome models denoted a series of independent parts with their specific physical properties (e.g.: interior-exterior, smooth-wrinkled, moduled-ornamented, regular-irregular) and their translation to the digital world to understand the small things that make up a whole ('thingness'), but also a three-dimensional understanding of the whole as a 'thicknesses' made up of small parts. The result of this process was an *analogue artifact*.

5 *ibidem*

6 Bergson, H. (1896) "*Of the Selection of Images for Conscious Presentation. What Our Body Means and Does*". Translated by N.M. Paul, N.M., Palmer, W.S. (1911) "*Matter and Memory*". London: George Allen and Unwin, p 8.

7 *ibidem*

DIGITAL TRANSLATIONS

DANIELA ATENCIO

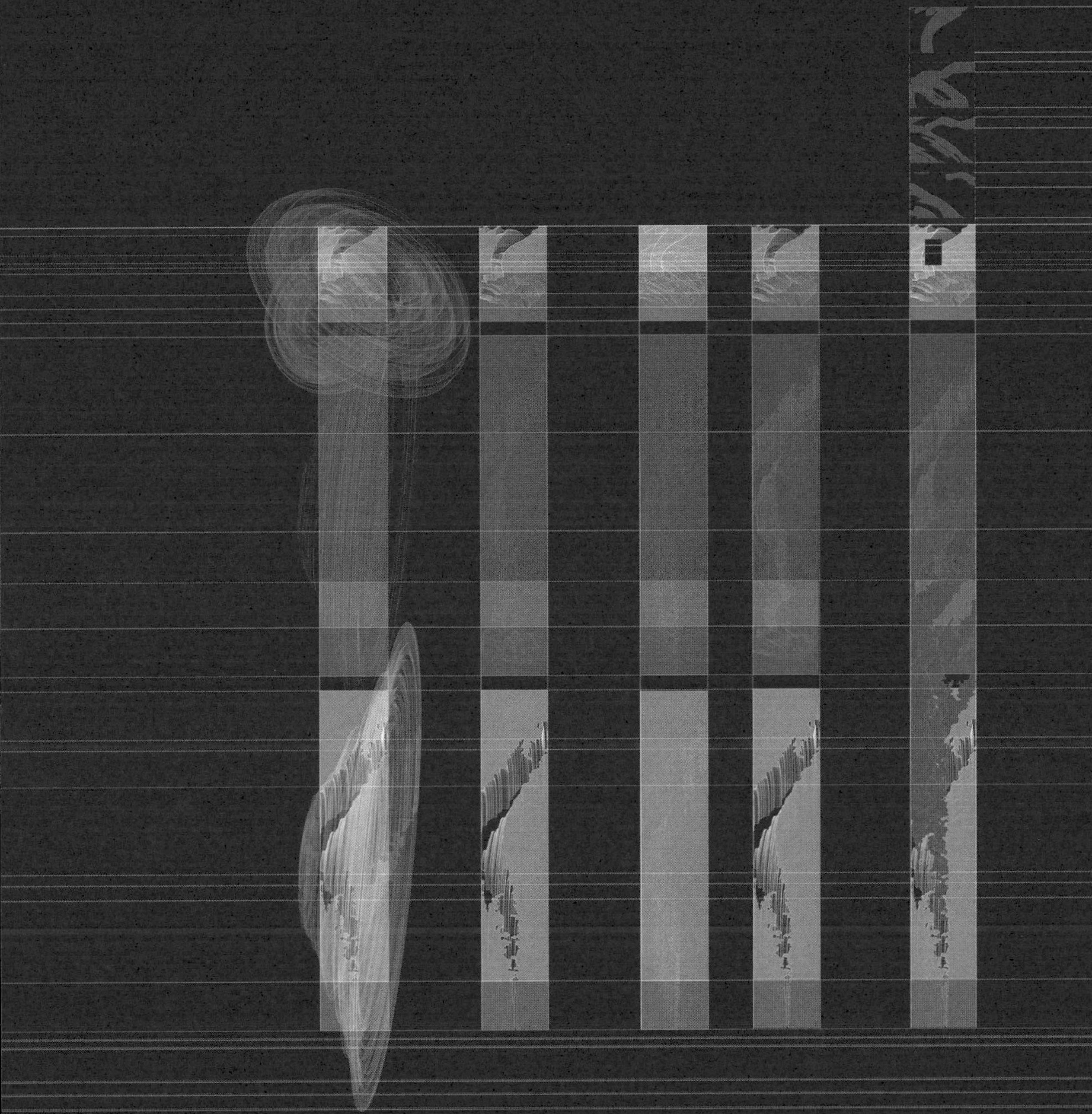

DIGITAL TRANSLATIONS DIAGRAM IN WORM-EYE-OBLIQUE VIEW. ABOVE: ROBOTIC TRANSLATIONS-RUINS. CENTER: OVERALL TRANSLATIONS. BELOW: ROBOTIC TRANSLATIONS-LANDSCAPES.

This translation was based on the conceptual findings and revealed material thickness conditions in the studied cases. It translates images and explores new readings into an abstract digital block: the *digital artifact*. This artifact and its set of iterations are then understood as an abstract, informed compression of digital material arrangements and capacities.

The *digital translation* had two design processes. The first, is the creation of the *digital artifact* that is constituted of two levels of information: On the one hand, the digital assembly of the parts, elements and material arrangements that come from the anatomical study of the cases in the previous phase. On the other hand, the material capacities that have to do with the most essential qualities or attributes that are expressed in the different environmental, social, political, or cultural abrasions and appropriations of the studied cases. The second design process translated the conditions explored before, with the use of the 6-axis robotic arm[1] toolpaths —in its first appearance as a 'bridge' between the digital artifact and the physical edition— in its movement to physical operations in the following phase of this workflow.

The robot, then, is the infrastructure ('bridge') and the tool that provided the tension between the different translations as design processes. It had three fundamental appearances in the workflow: as a translator between the digital artifact and the physical artifact, through the simulations and toolpaths developed in this chapter. As an editor of the physical artifact and, finally, as a translator between the physical world and the digital world, from the choreographic capture and the 3D scanning, with a different set of toolpaths. If we look back at these exercises, everything started with a set of images as the trigger of the whole process of translations (*analog translation*) and ends with another set of images from the final assembly of the *hyper-artifact*.

Going back to the creation of the translating experiment in the digital 3Dmodel, the fundamental concept is that it must be connected to the hermeneutics of the studied cases. It started with rereading the *analogic artifact* to accumulate, through digital addition and subtraction operations, different explorative iterations of the *digital artifact*.

These operations also tested, at the same time, compositional arguments (e.g.: stack, superpose, stratify, among others). For this reason, exploring each case clarified systemic data from the geometry and relationships of its parts (anatomical body). Properties such as materiality, abrasion, micro-organic life, patterns, and/or ornamentations were transferred within this translation through the algorithmic abstraction of 2D image vectors using scripts.

Images were remapped onto the surface by snapping the multiple generated vertices. Vertices were joined together with a continuous polyline and then converted to a boundary or surface; — the surface was then subdivided into an equal number of *UV*s. Extruding the surfaces simulated the assumption by which the case was constructed and how it was made (e.g., exterior-interior, interior- exterior, atmospherically, 360 degrees, by layers, among others). The outcome was a geometrical mixed-resolution digital block as a hybrid digital composition in terms of geometries and information from history, landscape, environmental, socio-cultural, and political readings using technological processes. Geometric complexity and mixed resolutions in the same object were converted into a 'total' digital matter, in other words, a *digital artifact*.

The second design process started form the use of a specific robotic software used to generate toolpaths to recreate the *digital artifact* geometries and its formal, spatial, and/or conceptual properties. The adapter and platform were loaded into the software as well as the digital 3Dmodel, with the specific digital robotic arm library, for the *autopath* recognition prompt to make a fast test from the robotic simulation.

Up to this step, what had been generated is the conformation of a digital environment for the interaction of the robotic arm and the *digital artifact*. This stage relates to develop a series of

1 ABB® (*Asea Brown Boveri*) IRB 1600 6-axis Robot, 1.2m reach, 10kg weight payload.

toolpaths that can simulate the development of a physical object by the *robotic translation* (following translation of the workflow). This step was fundamental from two perspectives: The first one, from anticipating the material edition —that must be physically executed in the workshop. This implied predetermining the type of tools[2] that can be used and assembled to the robot and execute them in the virtual environment of the robotic software. The second, is to anticipate distances and possible material behaviors with the material selected for the robotic edition.

For this, an aluminum-designed tool attachment for supporting a 30,000 RPM Mototool (*Bosch® GGS 28 CE*) was designed, modeled, imported, and attached to the robot axis #6 perpendicularly and aligned to the normal axis. The tool's target was the endpoint (or the last vertex where the attached bit would be). The tool creates its *Tool Central Point* (TCP), weight (Kg), center of gravity (X,Y,Z mm), and moment of inertia (Kgm2) and aligns the values with targets. The new tool is created in the platform's library, and it became the robot's new termination of axis #6.

The vertices, axes, and faces of the imported geometry are detected and can be reached by the robot's tool limits. This process detected the geometry traversing them and created targets through continuous and successive movement. Each new target was conformed to the limits of each of the geometric details of the *digital artifact*. The targets presented discontinuous orientations – preserved or not (non-linear movement, non-linear editing, non-Cartesian), or normal aligned or not (linear movement corresponding to the Cartesian axes). This point was critical to determining the edition's typology —the reinterpretation was equal or unequal to the source. (*Digital artifact*)

The tool, followed by the robot, is located at each target to verify their position, continuity, and range. It defines the depth of the toolpath, adjusting target positions and rotations according to the drill bit to be used and the depth required to edit, considering the material selected. The targets created a new toolpath, and each target generated an instruction or procedure.

In this step, the instruction type defines each target's movement: linear (true to Cartesian axes) or joint (non-linear, non-Cartesian motion). Both types were tested, studied, and mixed for the object in the upcoming robotic edition phase. It was then synchronized with the software's virtual controller to generate a specific code for each toolpath test. The module's code enabled the toolpath simulation.

The outcome of this translation ended up in multiple toolpath simulations at various positions and depths (modifying the position and rotation in mm of the X, Y, and Z axes) but also corresponding to various types of instruction (linear or joint) as well as recognizing the orientations of the original axis normal or aligning them to the surfaces of the artifact to address the mixed-resolution material capacities of the *digital artifact*.

2 See *Toolkit* chapter

ROBOTIC [PHYSICAL] TRANSLATIONS

DANIELA ATENCIO

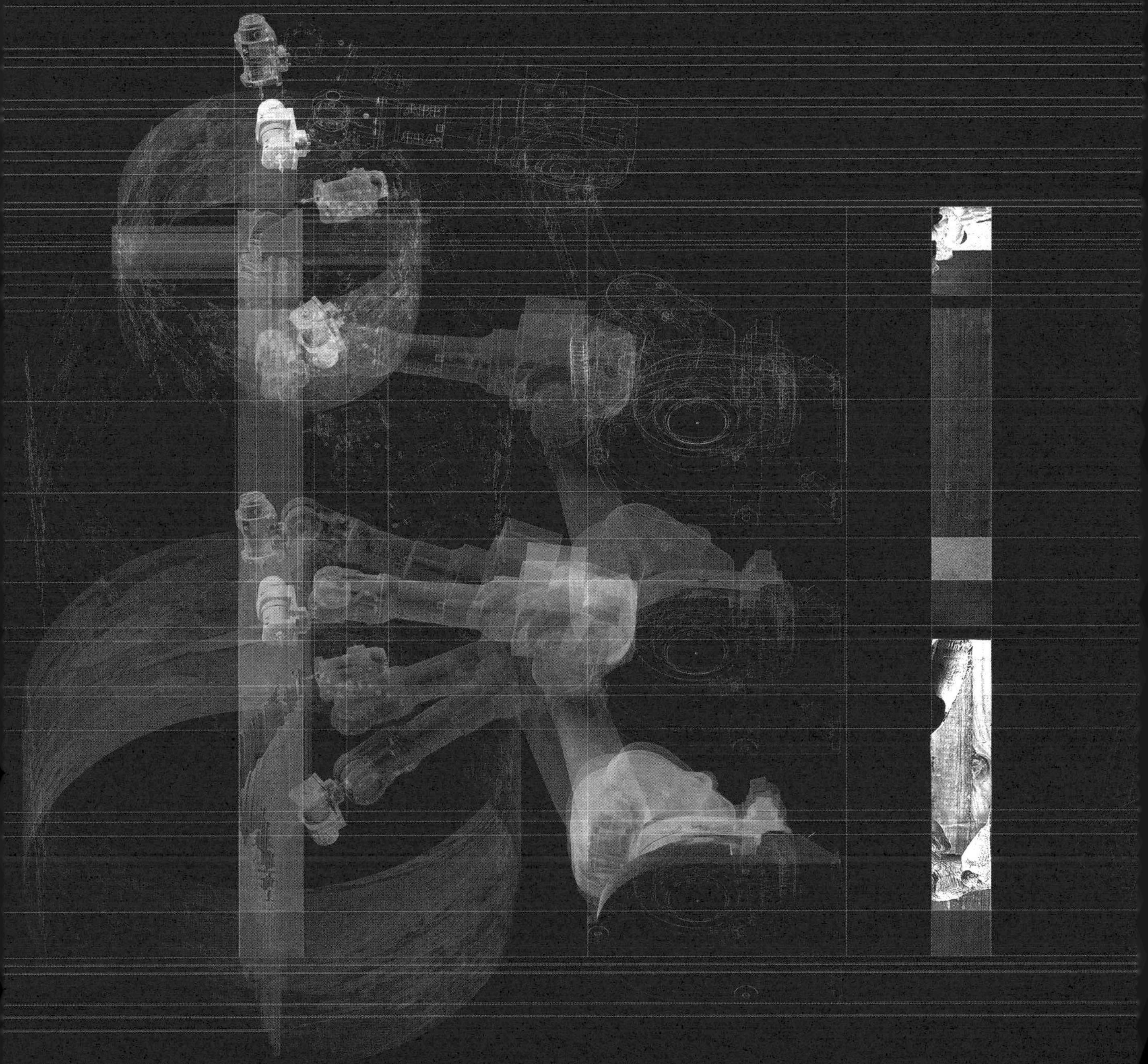

ROBOTIC [PHYSICAL] TRANSLATIONS DIAGRAM IN WORM-EYE-OBLIQUE VIEW. ABOVE: ROBOTIC TRANSLATIONS-RUINS. CENTER: OVERALL TRANSLATIONS. BELOW: ROBOTIC TRANSLATIONS-LANDSCAPES.

One of the most exciting approaches from the pedagogical perspective in the local context was editing some *physical blocks* in conjunction with the 6-axis robotic arm. This involved two processes: firstly, the translation of *digital artifacts* into physical objects by interpreting the geometries and effects of the toolpaths; secondly, the reconversion of the edited physical objects back into *digital artifacts*, using the choreographic movements of the robotic arm[1]. In both instances, toolpaths were the fundamental operations to execute these translations. Complementing this, it is important to remember that in the previous chapter was mentioned that the robotic translations were additionally deployed in the robotic simulation of the toolpaths[2]. For this part of the workflow, the focus was on the editing process of the physical objects.

The first procedures entailed the execution of the simulated toolpaths in the previous translation testing the real 6-axis arm robot space. The first task was the execution of the generated simulations into the surface of the physical artifacts (blocks), testing movements, distances, stability, or the first levels of edition in the material.

The blocks were made from layers of *Medium-Density Fiberboard* (*MDF*) to complete a set of different volumes (cubes or parallelepipeds) depending on the case and the iteration. The *physical MDF* blocks were selected and built for this process, after testing different types of woods or other dry laminar materials, because of its affordability and material density, that could give the students the possibility to execute, test, make mistakes, and test again, without almost any restriction or time constrains.

The blocks were built by laminar parts in layers of 18mm —conceptually related to the idea of layers that make up a whole of the studied and proposed *taxonomic artifact.* From this conceptual starting point, it's possible to speculate that the robotic arm and the execution of the toolpaths blurred the layers to edit a whole, erasing the lines that make up the cube and leading to a total new *physical artifact* via the robotic translation.

The same toolpath was repeated in several layers, calculating, and programming the depth, and mixing the tested drill bits in the previous translation, changing thicknesses or shapes: round, square, and triangular, among others. The first layers of edition were predictable and allowed the reading of the preprogrammed toolpaths. The challenge here was to start editing the surface and them going deeper without missing possible material mixed resolutions.

1 See *Inverse Translations*.

2 See *Digital Translations*.

After editing the surface, new material definitions emerged from subsequent layers. At this moment, a free-form editing process begun since the effects generated were unpredictable (material behavior).

The thicknesses of the Mototool drill bits and the editing layers from toolpaths, challenged the Cartesian axes using the specific capacities of the 6-axis robotic arm to perform different movements and joint behaviors. Unexpected actions of the drill rotation at 30,000 rpm, deterioration or friction on the material, and the geometry of the toolpath converted the simulation into micro-correspondence processes between material behavior and many imperfections in the execution of this translation.

The process was repeated in different faces of the MDF physical block, always trying to connect faces. Therefore, the material transformed from programmed paths, understanding the technical specificities (weight, density, mass, speed, among others) of the block and synchronizing the generated effects. This process was systematized and repeated in different iterations, generating different effects with dissimilar results. The transformations of the blocks suggested a new interpretation for the *physical artifact*.

It ceased to be a block and became a *robotic physical artifact*.

INVERSE [ROBOTIC] TRANSLATIONS

DANIELA ATENCIO

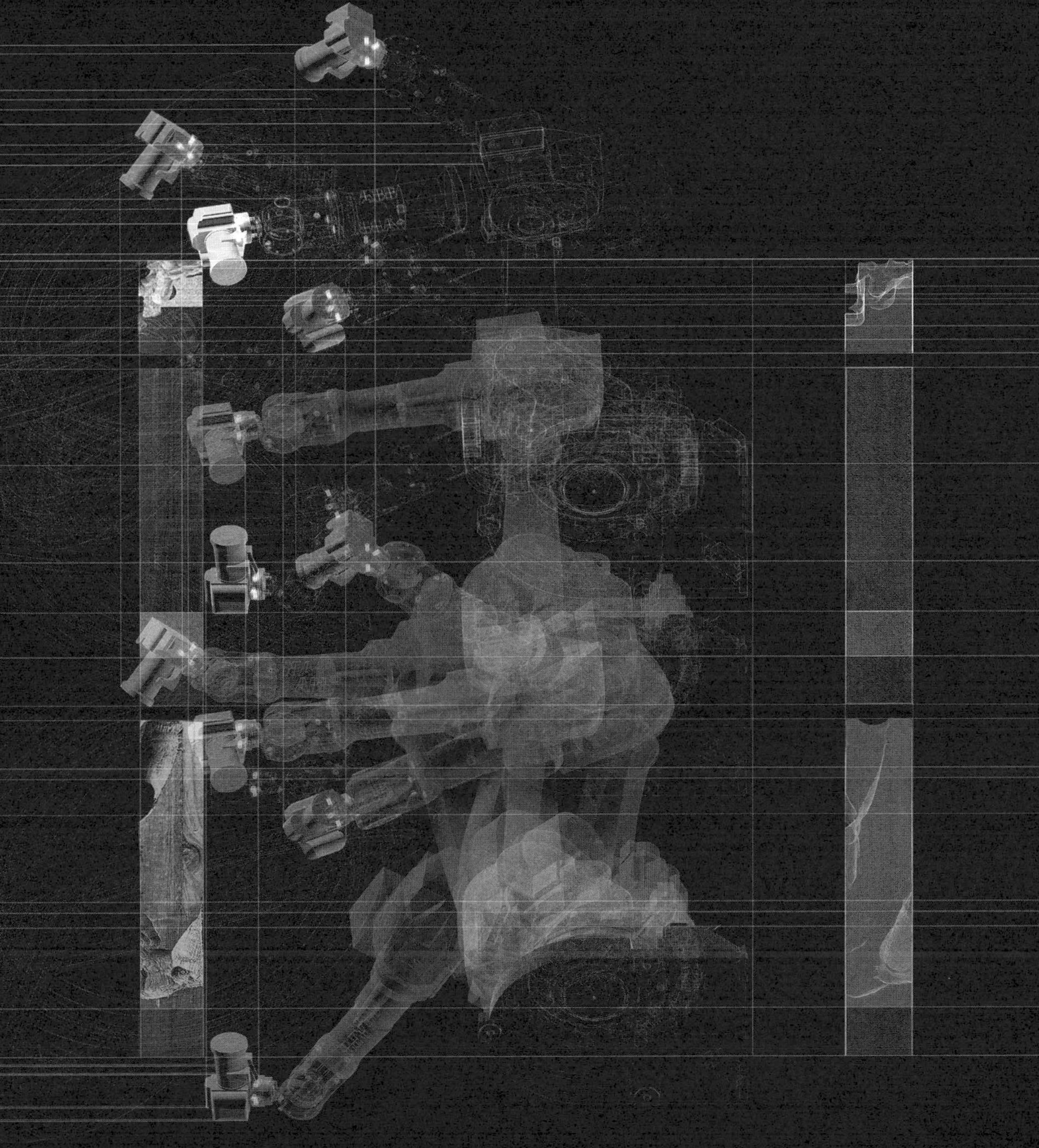

INVERSE [ROBOTIC] TRANSLATIONS DIAGRAM IN WORM-EYE-OBLIQUE VIEW. ABOVE: ROBOTIC TRANSLATIONS-RUINS. CENTER: OVERALL TRANSLATIONS. BELOW: ROBOTIC TRANSLATIONS-LANDSCAPES.

After having the final robotic physical objects, the second process was to return to the digital world through the choreographic record with the robotic arm. For this, a 3Dprinted tool was designed and created to attach a Nikkon® Z 14-30mm f/4 S camera to the robot axis #6.

The same path of the first part of the robotic interaction back in the digital translation were repeated to refer the camera tool to the digital platform and added to the library. In this case, the movements generated in the joint instruction allowed higher and smoother continuity. Following the physical geometry of the cubes, the targets permited the record-capture of geometries with greater detail. This process was called robotic choreography: a continuous movement with the camera and the robot, capturing the newly generated physical materiality.

The camera record generated a 3dmodel using pointcloud softwares. The precision of the robotic choreography allowed to generate a precise point cloud converted into a mesh to be later imported into a 3dmodel softwares with deep mapping projections. This registration gave the model multiple resolutions from the massive number of captured images with the robotic arm.

In other words, an exact reverse journey from physical to digital.

At this point, a parallel process occurred that had to do with the representation and recording of this whole pedagogical experiment. The observation of the process itself from the camera assembled to the robot axis #6, and the registration of the process from the outside (observing the observer).

In the first of the parallel processes, and using the expression of machine vision[1] , the interest was to have an objective designed observation. This, which seems anatomically and conceptually strange, promoted a series of debates during the process that were particularly nourishing for refining the experience with the robotic arm as a pedagogical tool, but on the other hand, it also fostered central discussions about representation.

Hardware and software are means, and as we have insisted since previous chapters, infrastructures and vehicles that allows the generation of design processes from different perspectives. What is interesting here is the power that takes not only the representation as part of the process, but also the agency given to the robot to build devices to represent these processes.

1 There were two ways of understanding '*machine vision*'. From what it was observed by the camera and the robot, to retrieve information and geometries (See: Lawrence Roberts Thesis (Ph. D.), Massachusetts Institute of Technology, Dept. of Electrical Engineering, 1963) and perhaps from the choreographic narrative of what was recorded closer to a cinematic dynamic representation.

When the robot moved around an object and records it with a camera that is attached on its axis #6, it can be said that it had a 'proper' and 'foreign' sight at the same time. Foreign because it is pre-established and designed by us from the toolpaths. It had predetermined shapes, movements, behaviors, and speeds. In this sense, it can also be established that the register of a design process can be predetermined, that is, represented. But it is also 'proper' because it had a series of mechanisms and micro mechanisms synchronizing themselves to generate the vision for which it was designed.

On the other hand, this process allowed to test the register of time. The robot is an infrastructure with movement or in other terms, a choreographic and expressive infrastructure, which runs through its spatial dome,[2] and its circumstances recalling Ortega & Gasset.[3]

2 Testa, P. (2017) "*Robot House*". New York: Thames & Hudson

3 Ortega y Gasset, J. ([1914] 1995) "*Meditaciones del Quijote*". (3.ª ed.) Madrid: Cátedra. See also the interview with Peter Testa in this book.

TRANSLATIONS ASSEMBLIES

DANIELA ATENCIO

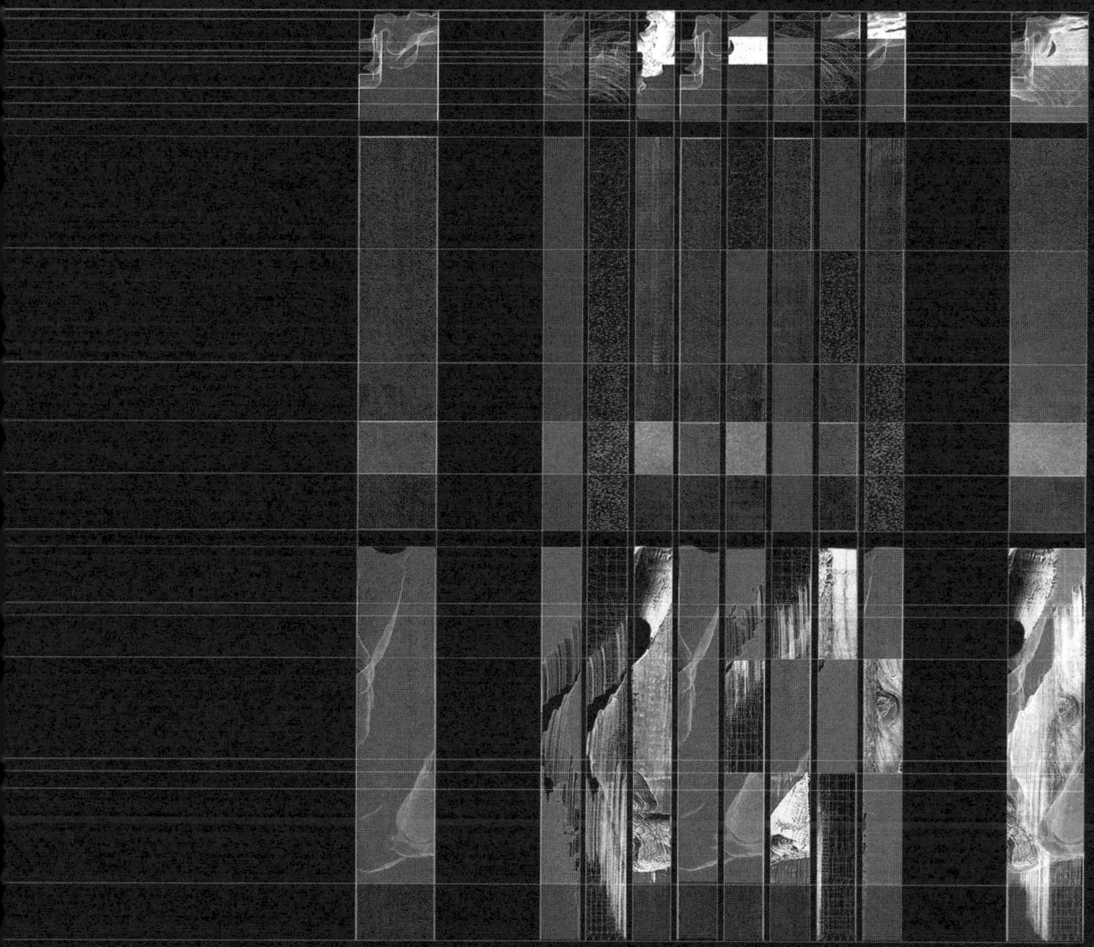

TRANSLATIONS ASSEMBLIES DIAGRAM IN WORM-EYE-OBLIQUE VIEW. ABOVE: ROBOTIC TRANSLATIONS-RUINS. CENTER: OVERALL TRANSLATIONS. BELOW: ROBOTIC TRANSLATIONS-LANDSCAPES.

Finally, the (re)construction or reinterpretation of all the previous translations must be understood as assembled entities that structured the *hyper-artifacts*. From the previously generated translations, the materialization of a hybrid representation or physical-digital render—digital and physical assemblage—was the result from this speculative exploration and the definitive argumentative accumulation of this investigation. The *hyper-artifact* carried multiple responsibilities and programmatic characterizations. The starting point of this specific translating process was to build and test the capacity in which the new artifacts hosted and materialized a broader and more ambitious discourse on the role of buildings in contemporary architecture.

The hyper-artifact is a hybrid architectural object generated through the reinterpretation of different features and multiple advanced technological and fabrication processes with the robot. The results or findings were the platform for the generation of programmed materials and the creation of new objects focusing on parts and processes based on data transformation. The condition of material complexity as the result of the superposition of different processes promoted the exploration of multi-temporal spaces (e.g.: historically reinterpreted or contemporarily ruined), unlocking new material capacities that were rethought, reprogrammed, and speculated. These artifacts were hyper-informed, hyper-temporal, synchronic, and diachronic.

Ultimately, the *hyper-artifacts* were the central experimental work of this research-creation, where images, objects, and processes were superimposed, and a series of translations allowed the discovery of work processes, performativities, and conceptualizations, always with technological mediation, computational thinking, and robotic manufacturing as transitional agents between the physical and the digital. The *hyper-artifacts* emerged from the rereading the studied cases to approach questions about the future, constantly questioning processes and mechanisms of design [1].

The *hyper-artifacts* were then a reconstruction of different parts, informed by the mediation of the robot. The rules of these hybrid assemblages, although seemingly focused on speculation, were built from two fundamental prerogatives: As objects informed primarily from the three-dimensional register of the *robotic artifacts*, and as coherently and consistent apparatus connected with the conceptual arguments observed in the *analog translation* (*taxonomic artifact*) and developed further in the digital artifact of the *digital translation*.

A final effect of this journey through the workflow was the challenge of representing the *hyper-artifact* through images. The whole process started with images of the case studies,

1 Bergson, H. (1896) "*Of the Selection of Images for Conscious Presentation. What Our Body Means and Does*". Translated by N.M. Paul, N.M., Palmer, W.S. (1911) "*Matter and Memory*". London: George Allen and Unwin, p 8.

from which a lot of information was obtained from the analysis and detailed observation. This workflow closed its path with images that demonstrate those capacities, but from the *hyper-artifact*. In short, and remembering *Testa*, "*images construct objects, and objects construct images*".[2] The final translation of this sentence could be images constructed artifacts; artifacts constructed *hyper-artifacts*, and *hyper-artifacts* constructed new images.

As a culmination of the workflow, these new images can be the loop that feeds back into the previous phases, becoming the input that could again incite a set of design processes understood from the *translations*. Also, the evaluation or measurement of the results, which in this case were exploratory and part of other incentives both pedagogical from the teaching of design processes in architecture, and transversal problems such as representation, leaded to the analysis of the *hyper-artifacts* located in the contexts from where they were born. In other words, there is also an analytical and purposeful loop that demonstrated that *hyper-artifacts* were born from contextual issues and returned to enhance the capacities of these places, no matter if they were built as historical objects or affected landscapes.

Both instances were programmable, or rather, *translated*.

2 Testa, P. (2017) "*Robot House*". New York: Thames & Hudson

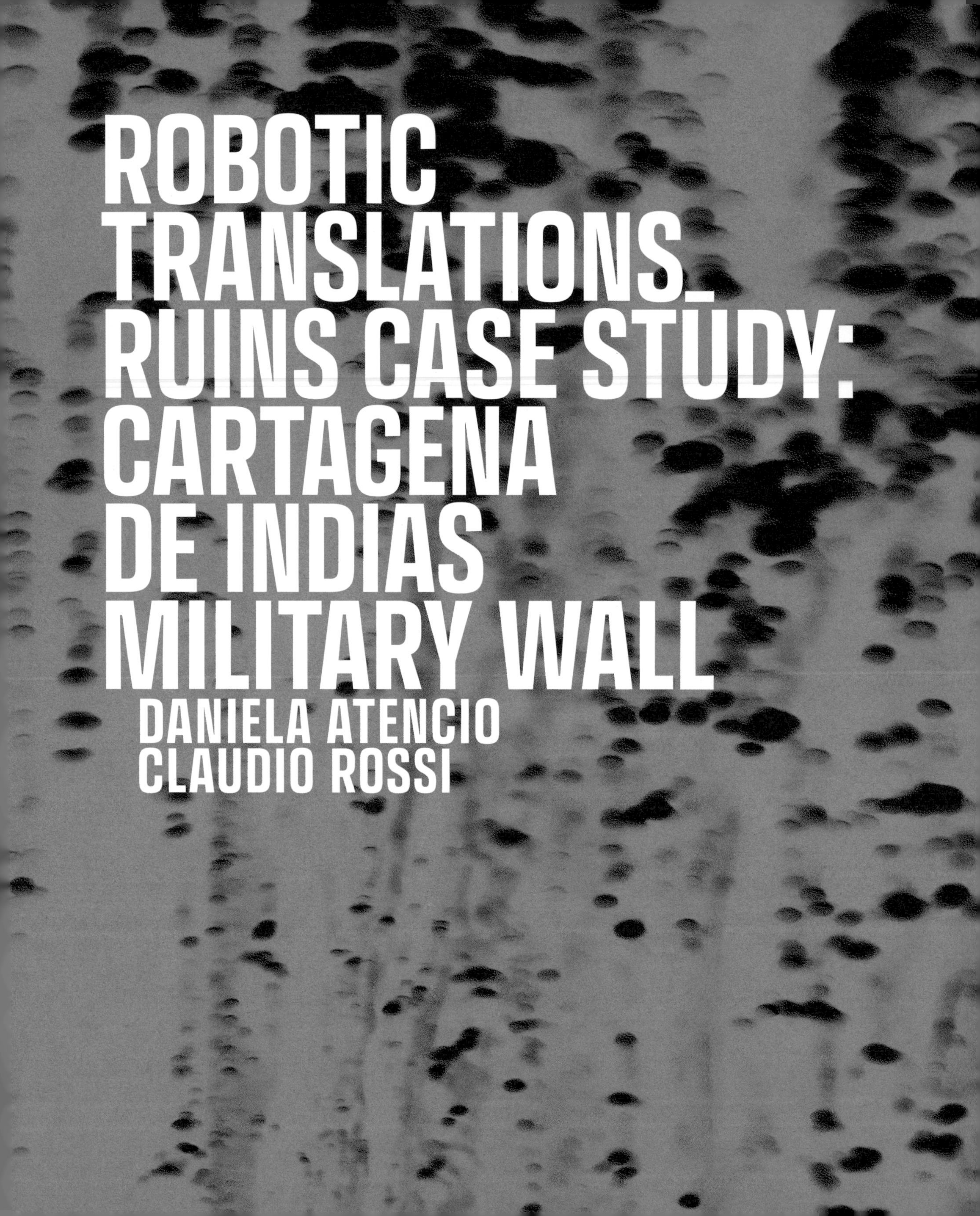

ROBOTIC TRANSLATIONS_ RUINS CASE STUDY: CARTAGENA DE INDIAS MILITARY WALL

DANIELA ATENCIO
CLAUDIO ROSSI

Material thickness, in any of its dimensions—constructive, structural, or formal—and affected by the factors of time, persistence, and/or climate, led to a fresh observation of some historical artifacts. One of the first study cases explored in the process of research-creation presented in this book was the military wall of Cartagena de Indias on the Colombian Caribbean coast. This piece of Renaissance-age infrastructure, although handled by UNESCO World Cultural and Historical Heritage since 1984, can be considered an artifact in ruination, with its temporal and historical beauty perpetrating new emerging material characteristics. The deterioration and destruction caused by time, salinity, and sea winds, the onslaught of Caribbean currents, humidity, and heat, among many factors, as well as the friction with daily public and social use, have turned the wall into a new great barrier with microorganisms embedded within the ancient coralline bricks.

The wall is a ruined infrastructure in a process of 'reconstruction' that, in a few hundreds of years, could become a reef on the shores of the Caribbean Sea. In which period should we place its material capacities? Is it in the immediate present or the future? It all leans, as may seem obvious, on a question of time. It is under construction because, although it is true that its material is in a constant state of decay, it is also giving way to new inhabitants and new purposes different from those for which it was built.

From these questions, the temporal focus of the architectural proposals, from its material determinations, connects with *Guilles Clement* positions in *Le Jardin en Mouvemont*. The author states that: "*the constructions of a man enter into an irreversible degradation process. Its inability to evolve condemns it, sooner or later, to ruin (...). On the contrary, nature never concludes anything.*"[1]. This biological dimension recalls the 'thick' materiality that Bruno Latour[2] interrogates, where the description enriched by the assembly of entities allows the observer to delve into the opacity of artifacts. This opacity is part of the anatomical constitution of *hyper-artifacts*; it carries the hidden responsibilities of the entities that can be studied, unlocked, and revealed from the hybrid technological approaches proposed in this research-creation.

This infrastructure, designed and built to protect one of the most important cities of the colonial circuit in Latin America from the Spanish crown, protected a key point between continental convergences—rivers, proximity to the Isthmus of Panama, among other conditions—and the Caribbean Sea. It is also part of the walled city system of the Caribbe-

1 Clément, G. (1991) "*Le Jardin en Mouvement*". Paris: Pandora.

2 Latour, B. (2007) "*Can We Get Our Materialism Back, Please?*", ISIS, 98(1), 138–142.

an Sea[3]. giving it a significant value for exploration: The topics reviewed and why we did not approach Cartagena can be transferred to other regions of Latin America, especially the islands or Caribbean.

The interest in taking this infrastructure was also due to its continental value. Inland and behind the wall, there are a series of landscapes and fresh or mixed water systems that give an ecosystemic richness to this military apparatus. Additionally, it is a great tourist magnet, which establishes a social anchorage between the local micro-economies of those who live inland and the different migrants who pass through this place. In other words, it is a social anchorage and a nest of human and non-human migrants.

From the steps proposed by the workflow and remembering the translations, the first design process developed was the taxonomic understanding of the wall. This observation and recording exercise was conducted on three scales: On the one hand, the characterization in sections comes from the reading of the historical object itself and its typological components. The sections were then established between bastions as the most robust and articulating pieces of the wall system. Based on these sections, the second reading and the quartering of other elements began: gates, sentry boxes, and folders, among others. The third step was the material quartering of all the constituent parts of these elements: vaults, domes, alfajías, parapets, and lintels, among many others. In this sense, the change in look allowed us to review not only the materiality of each part but also the system that joins, links, or repeats the different joints. Also to review, with detailed observations from the taxonomy itself, the understanding of small parts, the description of processes of deterioration and physical wear, or the review of possible new capacities lodged between their cries or erosions (e.g.: biofilm, among others).

From this, the transfer to the next phases of the workflow, starting with the digital translation, became an exercise of projectual and compositional exploration, but also conceptual and didactic. The wall demonstrated that built problems like these represent an opportunity to move a contemporary debate on recurring environmental and social issues in the region.

The digital translation accommodated the compression of much of the information observed at its different scales to, again, inform a design process that involves a series of transferees and learnings (and apprehensions) at each stage to inform the next. These tests were taken for physical exercise with the use of a robotic arm. The different blocks edited in this process put in value the possibility of repeating arrangements, edits, or material effects in

3 Paolini, R., & Gutiérrez, R. (1994) "*El Caribe Foritifcado*". Bogotá: Ediciones Uniandes.

a systematic and controlled way for the mass itself. Then, from the reverse journey, transfer these explorations to the hyper-artefacts that, according to the brief proposed by the workshop, should add to the wall new capabilities from the same environmental and social processes mentioned above.

One of the results, selected to show and refer to this exercise executed in Robot-Lab[4], as the academic space that connected the purposes of the research-programmed materialism with the pedagogical exercise of teaching the architectural project, developed a series of pieces that were placed outside the wall to extend its capabilities to the adjacent swamps. In other words, a series of hyper-artifacts informed by the wall but erected as new pieces extending paths, biological anchors, and ecosystem services. The 'new' and old walls coexist to expand their material capacities and give each piece new life.

4 Robot-Lab was the academic space designed by Professors Daniela Atencio and Claudio Rossi since 2021, to execute most of the questions developed in this document, but mainly as a space about pedagogy in the teaching of the undergraduate and graduate architectural project, around and from the robotic arm as an actor within the design process. This space pursued the construction of debates from technology, but also as a diverse, plural and creative place to encourage questions about the present and the future. Thanks to this dynamic, which strengthened a culture of research and creation in architecture, Robot-Lab ceased to be a course and migrated to a more robust research-creation space, which formulates its own projects and develops work paths that multiply the ecosystem of advanced digital fabrication from the robot, with fields of representation and computational thinking, with pedagogy and teaching of architecture.

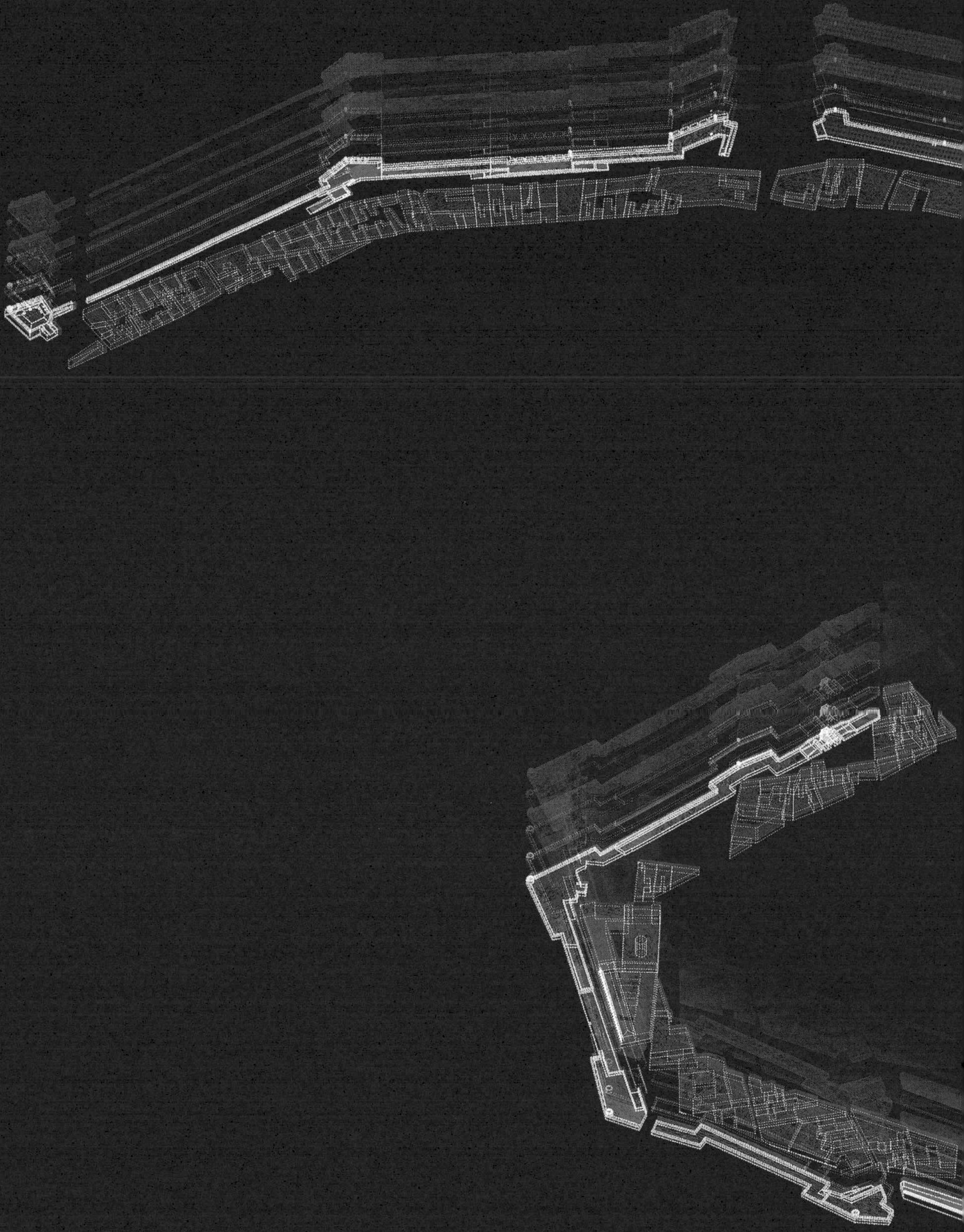

CARTAGENA DE INDIAS MILITARY WALL WORM-EYE VIEW DRAWING. EXISTING CONDITION [INCOMPLETE]. EXPLODED LAYERS [CONTEXT, STRUCTURE, MATERIALITY, MASS]

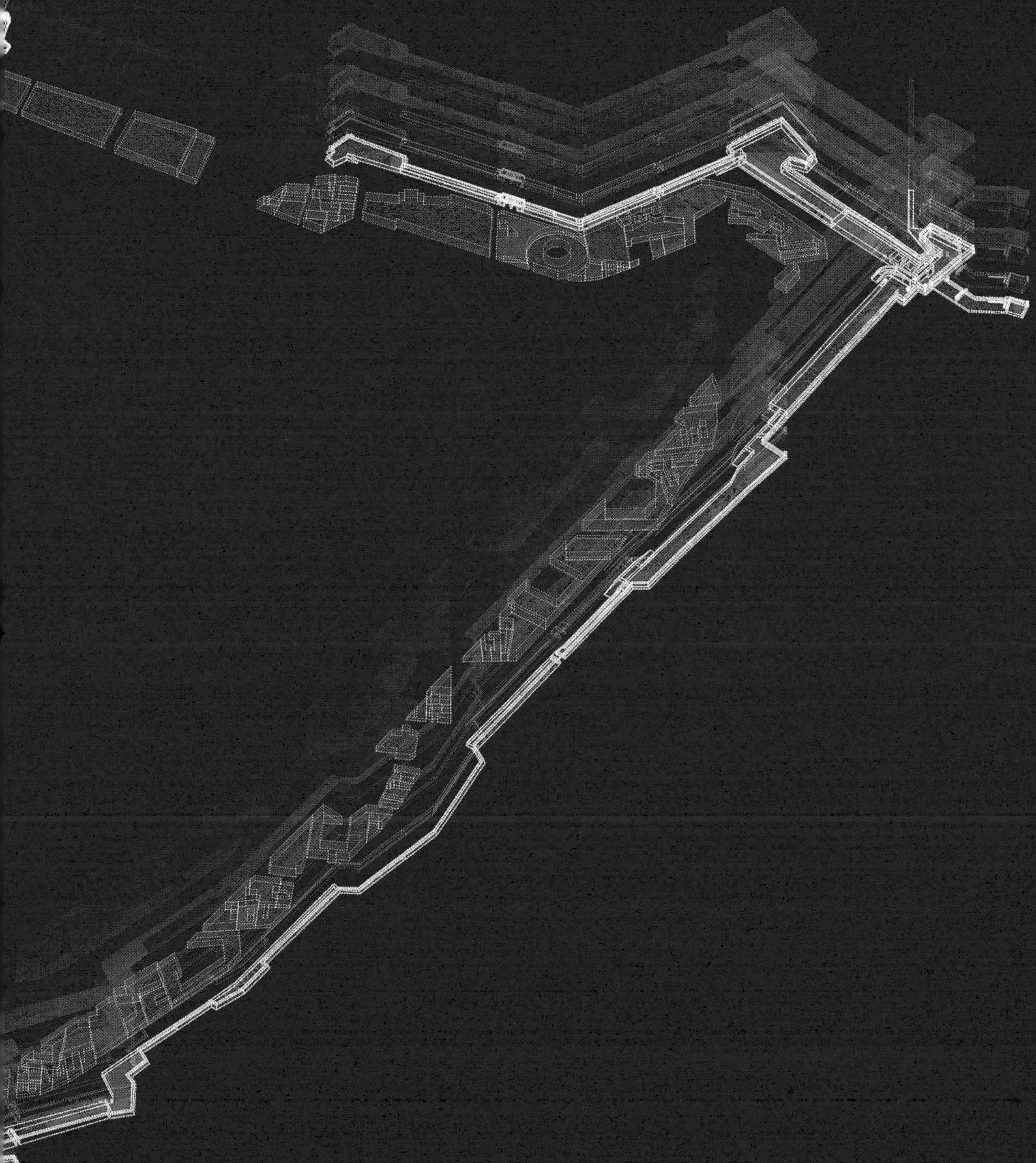

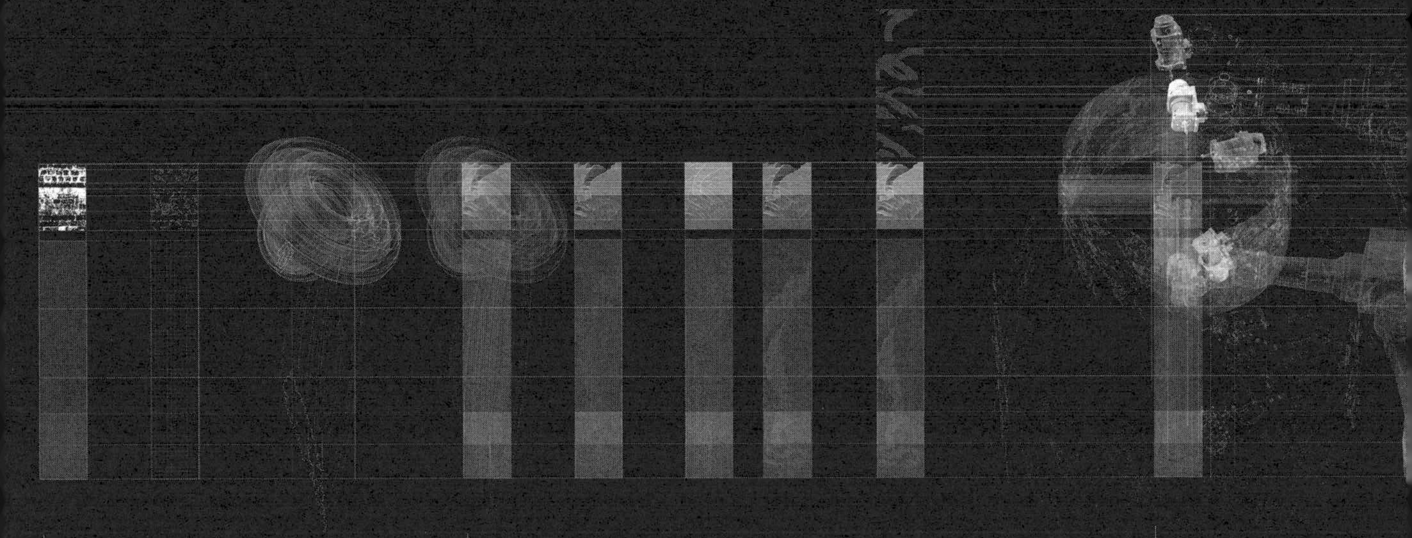

TRANSLATIONS WORKFLOW DIAGRAM IN WORM-EYE-OBLIQUE VIEW APPLIED TO ROBOTIC TRANSLATIONS-RUINS. CASE STUDY: CARTAGENA DE INDIAS MILITARY WALL FOR ROBOT_LAB COURSE 2022-10 AS A METHODOLOGICAL PROCESS.

ANALOGUE TRANSLATIONS

NEW MATERIALISM

DIGITAL TRANSLATIONS

UNLOCKING MATERIALITY

ROBOTIC TRANSLATIONS

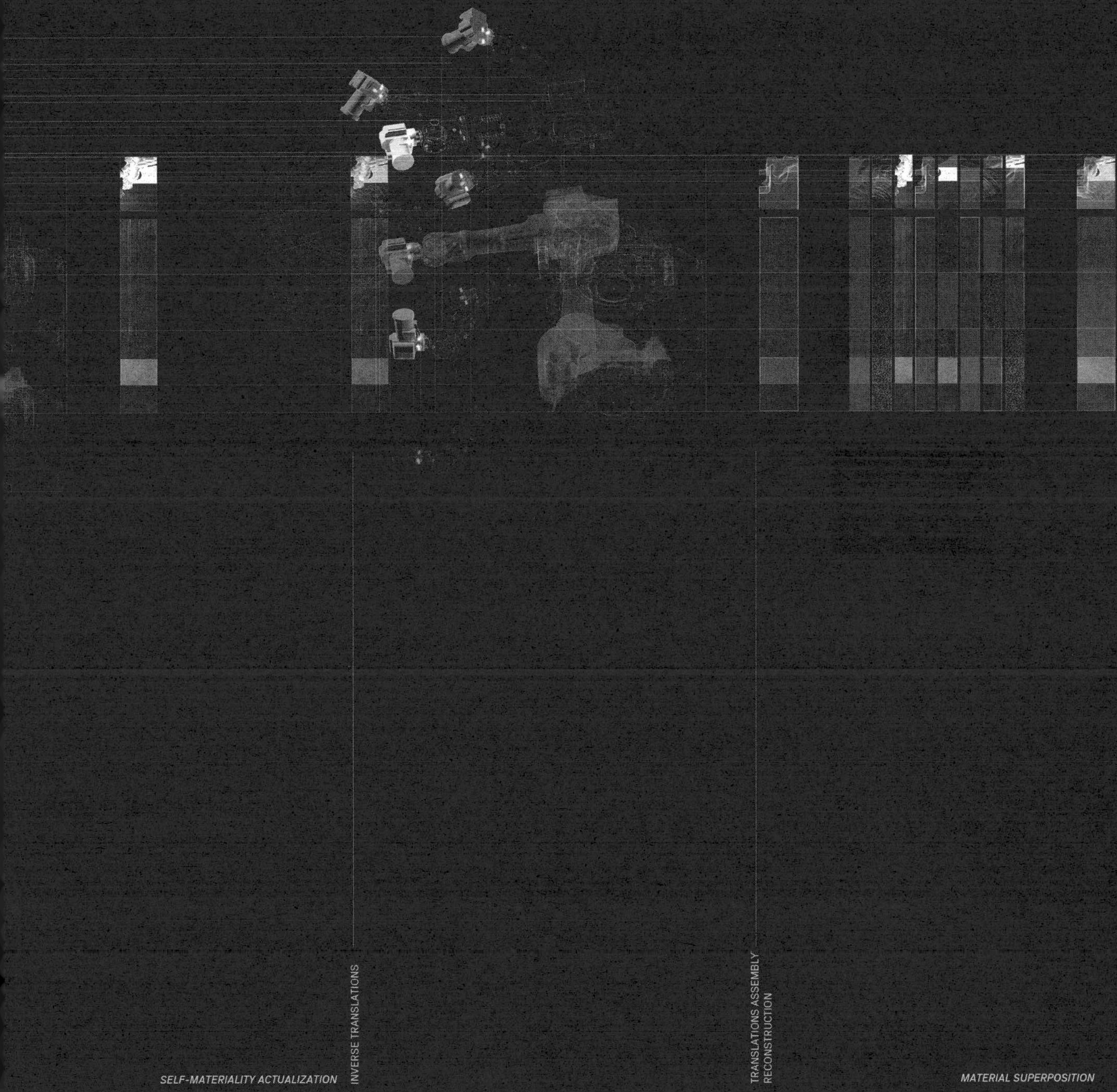
SELF-MATERIALITY ACTUALIZATION
INVERSE TRANSLATIONS
TRANSLATIONS ASSEMBLY
RECONSTRUCTION
MATERIAL SUPERPOSITION

ANALOGUE TRANSLATIONS - PEDAGOGICAL RESULTS
STUDENTS: LINA TRIVIÑO, SOFIA URIBE, JUAN MARTIN SÁNCHEZ.
FROM THE PROJECT: COMPLEXITY AND CONTRADICTION.
FACULTY-PROFESSORS: DANIELA ATENCIO, CLAUDIO ROSSI.
COURSE: ROBOT_LAB. UNIVERSIDAD DE LOS ANDES. BOGOTÁ, COLOMBIA. 2022

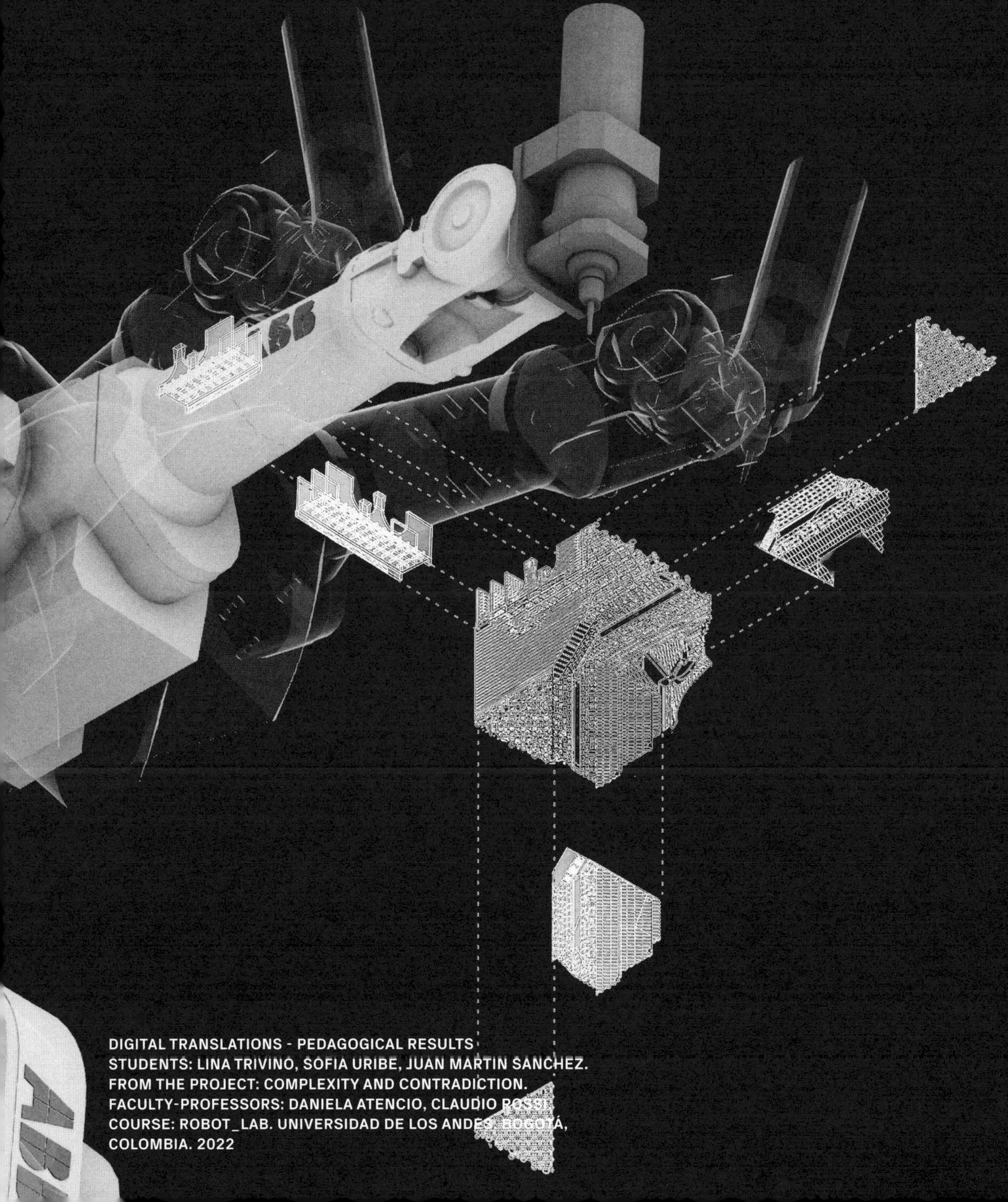

DIGITAL TRANSLATIONS - PEDAGOGICAL RESULTS
STUDENTS: LINA TRIVINO, SOFIA URIBE, JUAN MARTIN SANCHEZ.
FROM THE PROJECT: COMPLEXITY AND CONTRADICTION.
FACULTY-PROFESSORS: DANIELA ATENCIO, CLAUDIO ROSSI
COURSE: ROBOT_LAB. UNIVERSIDAD DE LOS ANDES. BOGOTÁ, COLOMBIA. 2022

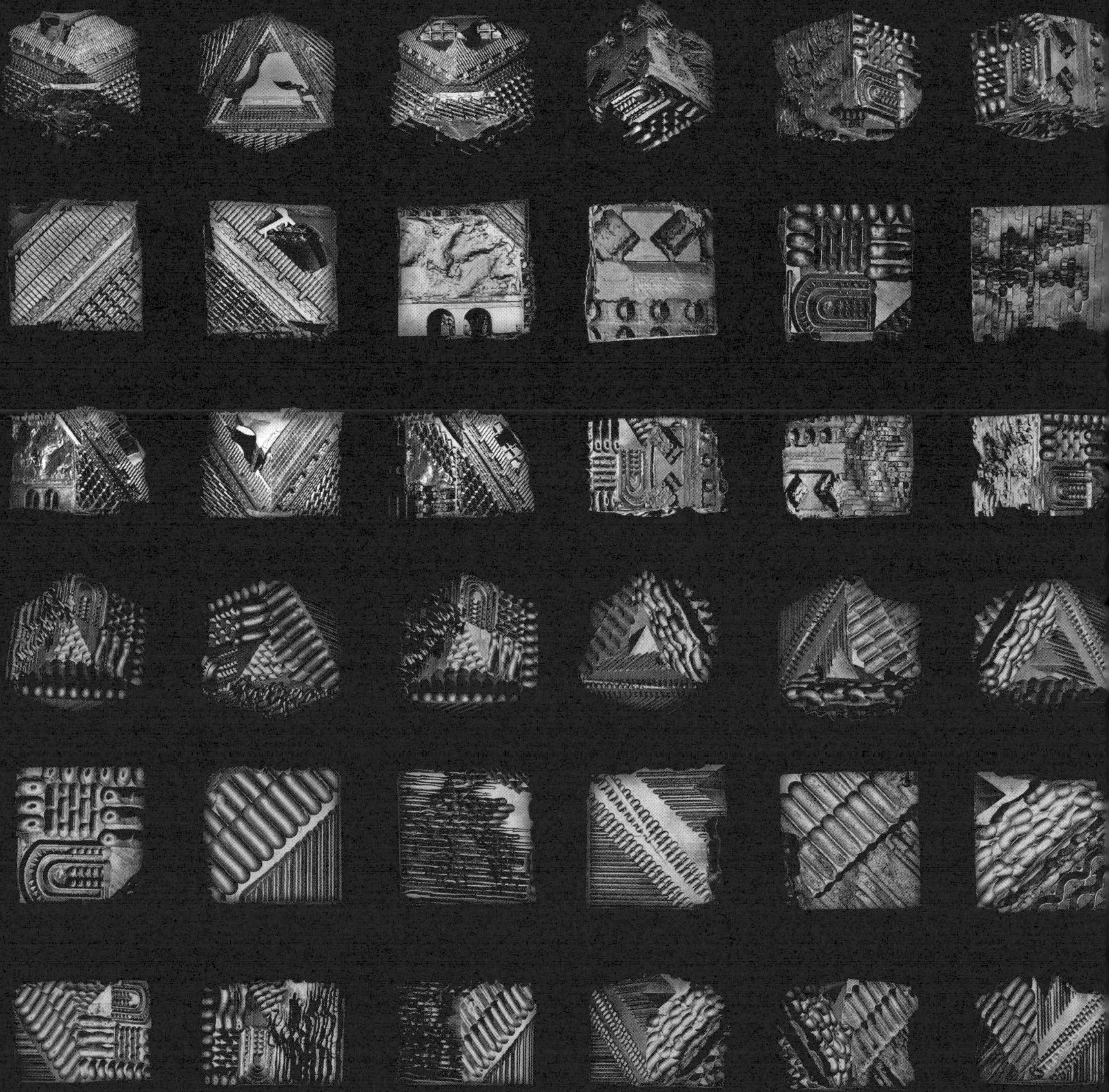

ROBOTIC [PHYSICAL] TRANSLATIONS - PEDAGOGICAL RESULTS
STUDENTS: LINA TRIVIÑO, SOFIA URIBE, JUAN MARTIN SÁNCHEZ.
FROM THE PROJECT: COMPLEXITY AND CONTRADICTION.
FACULTY-PROFESSORS: DANIELA ATENCIO, CLAUDIO ROSSI.
COURSE: ROBOT_LAB. UNIVERSIDAD DE LOS ANDES. BOGOTÁ, COLOMBIA. 2022

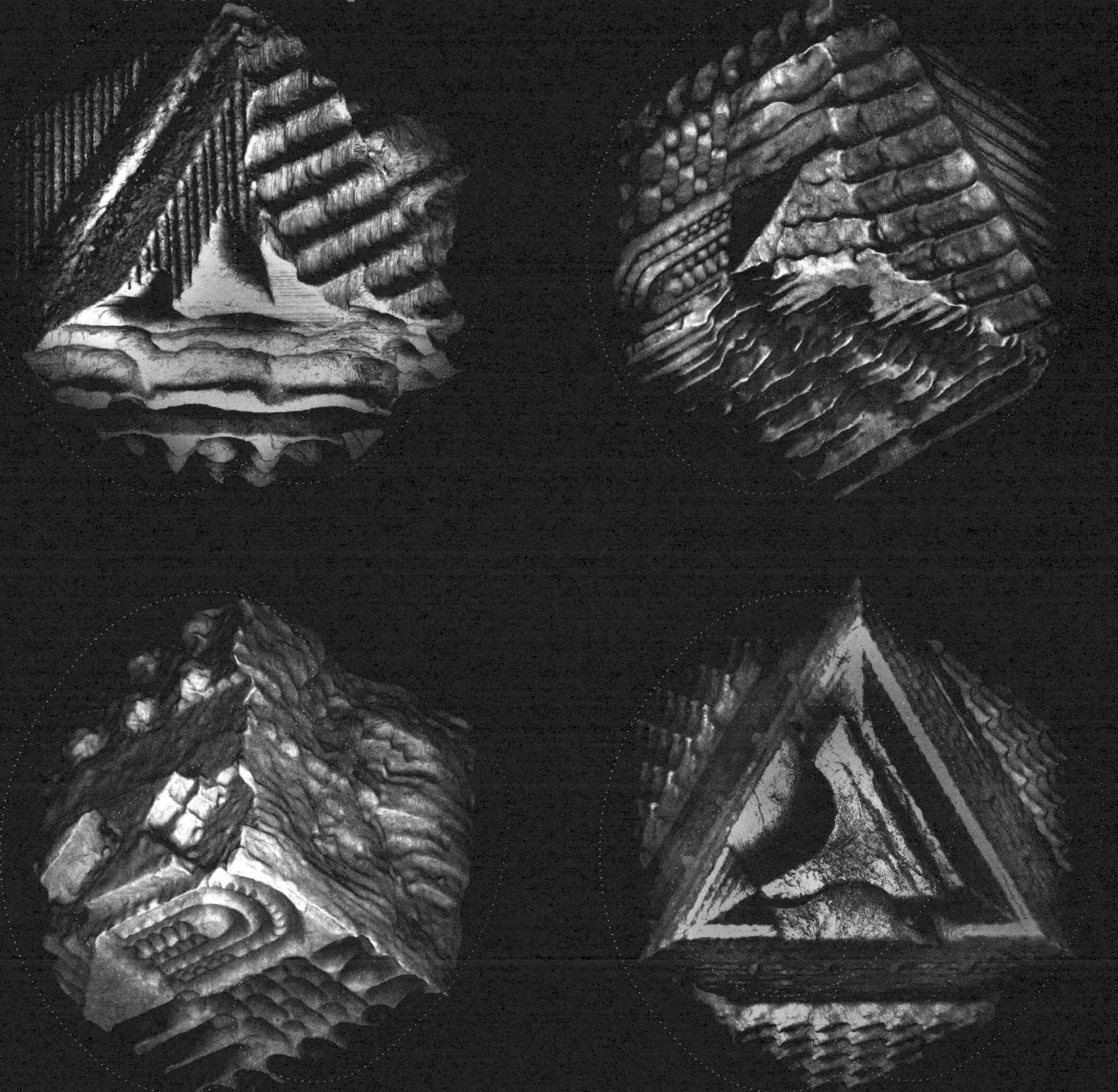

INVERSE [ROBOTIC] TRANSLATIONS - PEDAGOGICAL RESULTS
STUDENTS: LINA TRIVIÑO, SOFIA URIBE, JUAN MARTIN SÁNCHEZ.
FROM THE PROJECT: COMPLEXITY AND CONTRADICTION.
FACULTY-PROFESSORS: DANIELA ATENCIO, CLAUDIO ROSSI.
COURSE: ROBOT_LAB. UNIVERSIDAD DE LOS ANDES. BOGOTÁ, COLOMBIA. 2022

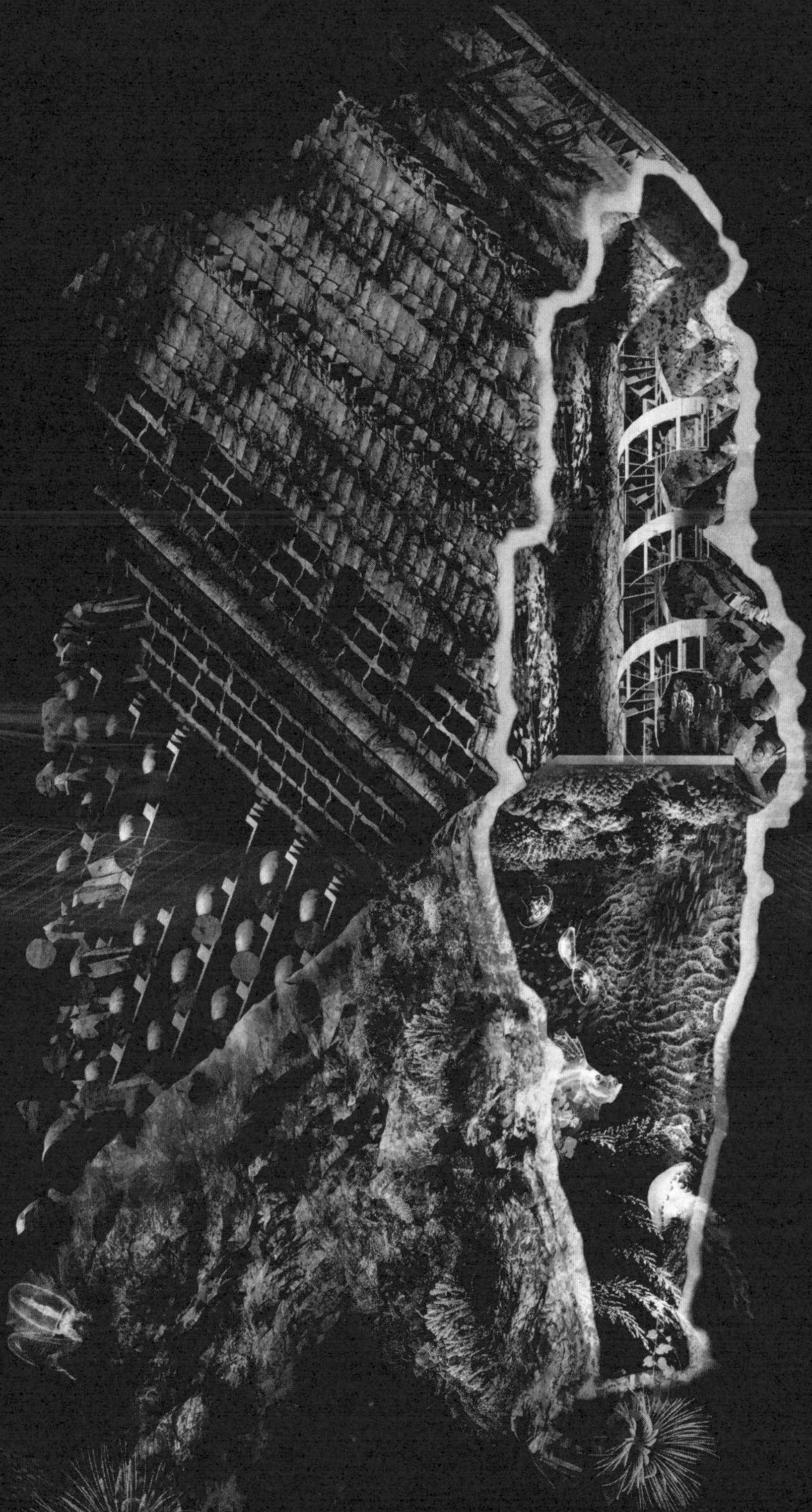

TRANSLATIONS ASSEMBLIES: HYPER-ARTIFACT - PEDAGOGICAL RESULTS _CHUNK 1. STUDENTS: LINA TRIVIÑO, SOFIA URIBE, JUAN MARTIN SÁNCHEZ. FROM THE PROJECT: COMPLEXITY AND CONTRADICTION. FACULTY-PROFESSORS: DANIELA ATENCIO, CLAUDIO ROSSI. COURSE: ROBOT_LAB. UNIVERSIDAD DE LOS ANDES. BOGOTÁ, COLOMBIA. 2022

TRANSLATIONS ASSEMBLIES: HYPER-ARTIFACT - PEDAGOGICAL RESULTS _CHUNK 2. STUDENTS: LINA TRIVIÑO, SOFIA URIBE, JUAN MARTIN SÁNCHEZ. FROM THE PROJECT: COMPLEXITY AND CONTRADICTION. FACULTY-PROFESSORS: DANIELA ATENCIO, CLAUDIO ROSSI. COURSE: ROBOT_LAB. UNIVERSIDAD DE LOS ANDES. BOGOTÁ, COLOMBIA. 2022

ROBOTIC TRANSLATIONS_ LANDSCAPES CASE STUDY: LA GUAJIRA DESERT

DANIELA ATENCIO
CLAUDIO ROSSI

Hyper-artifacts are, as explained in previous chapters, a mix between buildings and machines. Buildings with specific materialities, programs, and spatial definitions, but also as infrastructures that endow the diverse landscapes where they operate with capacities; and machines, which are inserted in a particular nature and with specific environmental responsibilities. In this case, we hope that they can become explorations for the selected context: La Guajira.

From the conformation of the northern strip of Chile in Atacama and Antofagasta to the dunes of Coro in Falcón-Venezuela, we can register desert soils that build a singular landscape within the green image of our continent. *Richard Weller* in the *Atlas for the End of the World*[1] proposes to look at what he calls: Biological Hotspots. These places, located in different parts of the world, represent the opportunity to recompose certain processes of climate erosion, but also to reverse, from resilience and landscape, environmental phenomena, and processes necessary to mitigate climate change thanks to anthropogenic impact. In other words, these places are and will be the geographical spaces from where we can rebuild the world and why not, as *Weller* proposes, save it.

Paradoxically, this same research has determined that a large part of the possible places of projectual attention are precisely accumulated in the Global South and many of them in Latin America, as spaces where shortages, inequalities, and/or corruption are added as part of their daily logics.

One of these Biological Hotspots is contained in the Caribbean coast of Colombia and within them the northeastern extension in the department of La Guajira. This exploration focused its attention on this geographic space, particularly on its desert extension as a historical region[2], which joins the north of Zulia State in Venezuela to the west of Manaure in the Department of La Guajira, Colombia.

This geographic position demarcates a series of lessons from the reading of limits, borders, and frontiers as initial concepts.[3] La Guajira expands within the different landscape strips and diverse activities of multiple scales. From artisanal coal mining to the mega-exploitation of Cerrejón. From artisanal salt collection to the large industrial salt lagoons of Manaure. Rites associated with the ways of living in the desert by the *Wayuu*[4] community, and all its logics or wisdoms. As *Weller* warns in his cartographies, La Guajira is one of the Departments with the most challenges in the region, from malnutrition[5] and access to resources such as water and energy. However, these characteristics are contrasted with the ability to adapt to multiple situations and be resilient communities.

The *Yonna*[6], the weavings, the women water carriers, the artisanal salt, continue to be practices of great cultural and material richness. The border and the desert apparently form an indissoluble relationship. The transitivity between common geographical spaces but divided by administrative or political boundaries such as countries, contradict the definitions or spatial wills of these places. This is what *Germán Cardozo Galué* has called: *Histori-*

1 Weller, R., Hoch, C., & Huang, C. (2017) "*Atlas for the End of the World*". Philadelphia: University of Pennsylvania. http://atlas-for-the-end-of-the-world.com

2 Cardozo Galué, G. (2013) "*Significaciones y Alcances del Concepto Región Histórica*". Cambios & Permanencias (4): 45 – 61.

3 Rossi C. (2021) "*Límites, Bordes y Fronteras*". In: "*Grandes Lecciones, Pequeñas Edificaciones*". Bogotá: Ediciones Uniandes, pp. 60-69.

4 The word *Wuayuu* means 'person' in *Wuayuunaiki*, the language of the people of the Guajira Desert who settled in this region approximately 150 BC.

5 Official malnutrition data in La Guajira are the second highest in Colombia. These reports can be officially reviewed in the weekly bulletin of the National Institute of Health. To understand this phenomenon, see: https://cerosetenta.uniandes.edu.co/tumbas-sin-nombre-los-ninos-mueren-por-desnutricion-en-la-guajira-y-nunca-son-reportados/

6 Ver: Carrasquero, A., Finol, J., & García, N. (2009) "*Símbolos, Espacio y Cuerpo en la Yonna Wayuu*". Revista de Ciencias Sociales, 15(4), 635-652.

cal Regions[7] as geographic spaces or regions whose past is contained and consolidated, transcending more contemporary logics such as political-administrative divisions. It is impossible no to refer to the northern Mexico and lower California or west Texas as a single space as studied by *Ronald Rael* [8], or *Tatiana Bilbao et. al.* [9], among others. Therefore, it is impossible to determine also a Colombian guajira from a Venezuelan one.

Lotman[10], dedicated a large part of his observations to the ways of inhabiting. His contributions have to do with the construction of definitions that normally contradict at least conventional logics. Among them, he speaks of the marvelous permeability of the habitable spaces of certain cultures or habitable interculturality. Definitions such as property, public or private are questioned from the different pieces that make up the occupations of the desert. The *Cují* (a tree characteristic of these places), by providing shade, becomes a habitable space for the desert, an informal classroom, or a communal meeting place. The 'ranchería', the 'chinchorro' and the Cují are fundamental parts of these logics.

We are focusing on a few authors to establish a series of project concerns. The first, the recent research by *Samia Henni, Deserts are Not Empty*[11], as the negation of the idea of deserts as empty places. The author establishes the colonial opposition to the idea of something 'desert' or 'empty' as a potential to be rethought from its own logics. Deserts are full of very particular inhabitants, materialities and spatial expressions that must be read and understood, and then converted into very special forms of action.

The second, *Watanabe*, Peruvian poet, and architect, from his phenomenological characterizations of the Peruvian desert in the *Piedra Alada* [12]. Such a description is filled with a wealth of details about materialities, atmospheric phenomena, human and non-human inhabitants, and their diverse ways of constructing these landscapes north of Lima (Moche Desert):

> *"Strangely enough*
> *one of its wings lingered on the stone's back,*
> *its gelatinous sinews dried out*
> *and* ***adhered to the stone***
> ***as if it were a body.***
> *For several days*
> *the sea wind*
> *uselessly beat the wing, it beat*
> *without understanding*
> *that we can imagine a bird, the most*
> *beautiful bird,*
> *but not make it fly."*[13]

Design decisions must be related to responsibilities, particularly in the most contemporary logics. These responsibilities must be adjusted to reality and context. In the case of the desert, and from what was mentioned at the beginning of this text, they must be understood as spatial, environmental, cultural, and social capacities; in other words, water, energy, shelter, and food, among others. From these aspects, and without forgetting all the data from different governmental and non-governmental organizations on the Guajira, we can understand that the specific issues associated with hyper-ar-

7 Cardozo Galué, G. *ibidem*

8 Rael, R. (2017) "*Borderwall as Architecture: A Manifesto for the U.S.-Mexico Boundary*". Berkeley: University of California Press.

9 Bilbao, T., Greenberg, N., Ghosh, A. (2020) "*Two Sides of the Border. Reimagining the Region*". Zurich: Lars Müller Publishers.

10 Lotman, J. (2005) "*On the Semiosphere*". Translated by Clark, W. Sign Systems Studies, 33.1

11 Henni S. (2022) "*Deserts are Not Empty*". New York: Columbia University Press.

12 Watanabe, J. (2005) "*La Piedra Alada*". Valencia: Editorial Pre-Textos.

13 *Ibidem*. Translation and underlining by the author, denoting the spatial and material character of the desert.

tifacts are approached through a very particular programmatic exploration.

The projects must challenge the programmatic conditions of the place, building hybrids that complement the different local deficiencies, as well as environmental ambitions as proposed by the Atlas of the End of the World. They are understood as buildings and machines (artifacts) whose programmatic base is heterotopic: schools- 'jagueyes' (*wuayuunaiki* ponds), health clinics—nurseries, or housing-salt cultivation—are illustrative examples of this logic. These are just a few examples of the inevitable relationship between infrastructure and landscape. An architecture that, regardless of its specific program, demonstrates responsibility and understands the temporality of the Wayuu context.

To achieve all this, the application of the workflow is fundamental. The first step, similar to the case study in Cartagena, is the taxonomic observation from three scales. The first is the selected desert space in its different landscape strips: the sea and its shore, the mangrove strip, the industrialized salt pools, the arid desert, and the dry tropical forest. For each strip, it was necessary to understand its material specificities and its relationships with adjacent strips. Finally, there is the materiality itself: salt, cracked soils, dryness, sediments, and so on.

Again, this observation was the trigger for the following translations. Two of the most interesting challenges from a pedagogical point of view were the incorporation of the robot as an infrastructure that allowed us to understand another scale of editing, that is, from the most conceptual basis, the opportunity to 'edit' with the robot landscapes.

The premise was then that, as can be seen in the anthropic effect of machines operating in the industrialization of salt and, therefore, modifying the soil and its material, environmental, cultural and landscape connotations, the robot can be a new agent in the 'programming' of these surfaces and ultimately of the desert landscape. This vision is reminiscent of the phrase: *"A society needed technological sophistication in order to transform the desert; untransformed"* from *Patricia Nelson Limerick* in *Deserts Passages* [14]

The result selected to demonstrate this exercise from our Robot_Lab course, was a hyper-artefact on the mangrove fringe on the desert coast of Manaure. This piece, informed by different material resolutions, opposes the horizontality of the place to stand as a testimony of environmental wear but also as a vertical infrastructure that understands the temperature changes and promotes a kind of vertical cistern to load the aquifers of the mangroves. A piece made of material capacity that, although it imposes itself in the landscape, is designed to be ruined and disappear once the mangrove population and its marshy soil are recovered.

14 Limerick, P. (1985) "*Desert Passages. Encounters with the American Deserts*". Albuquerque: University of New Mexico Press.

LA GUAJIRA DESERT [A STRIP OF MANAURE: SEA, MANGROVE, SALT INDUSTRY-POOLS, DESERT, TROPICAL-DRY FOREST]. WORM-EYE VIEW DRAWING. EXISTING CONDITION. EXPLODED LAYERS [GEOMETRY, TOPOGRAPHY, EROSION, VEGETATION, AMONG OTHERS]

LA GUAJIRA DESERT [A STRIP OF MANAURE: SEA, MANGROVE, SALT INDUSTRY-POOLS, DESERT, TROPICAL-DRY FOREST] PHYSICAL MODEL. FABRICATED [CNC, 3DPRINTING] FOR ROBOT_LAB COURSE 2023-10 AS A METHODOLOGICAL TOOL IN ITS EXISTING CONDITION. INTERVENED BY THE END OF THE TRANSLATIONS WORKFLOW WITH 3DPRINTED HYPER-ARTIFACTS.

TRANSLATIONS WORKFLOW DIAGRAM IN WORM-EYE-OBLIQUE VIEW. APPLIED TO ROBOTIC TRANSLATIONS-LANDSCAPES. CASE STUDY: LA GUAJIRA DESERT FOR ROBOT_LAB COURSE 2023-10 AS A METHODOLOGICAL PROCESS.

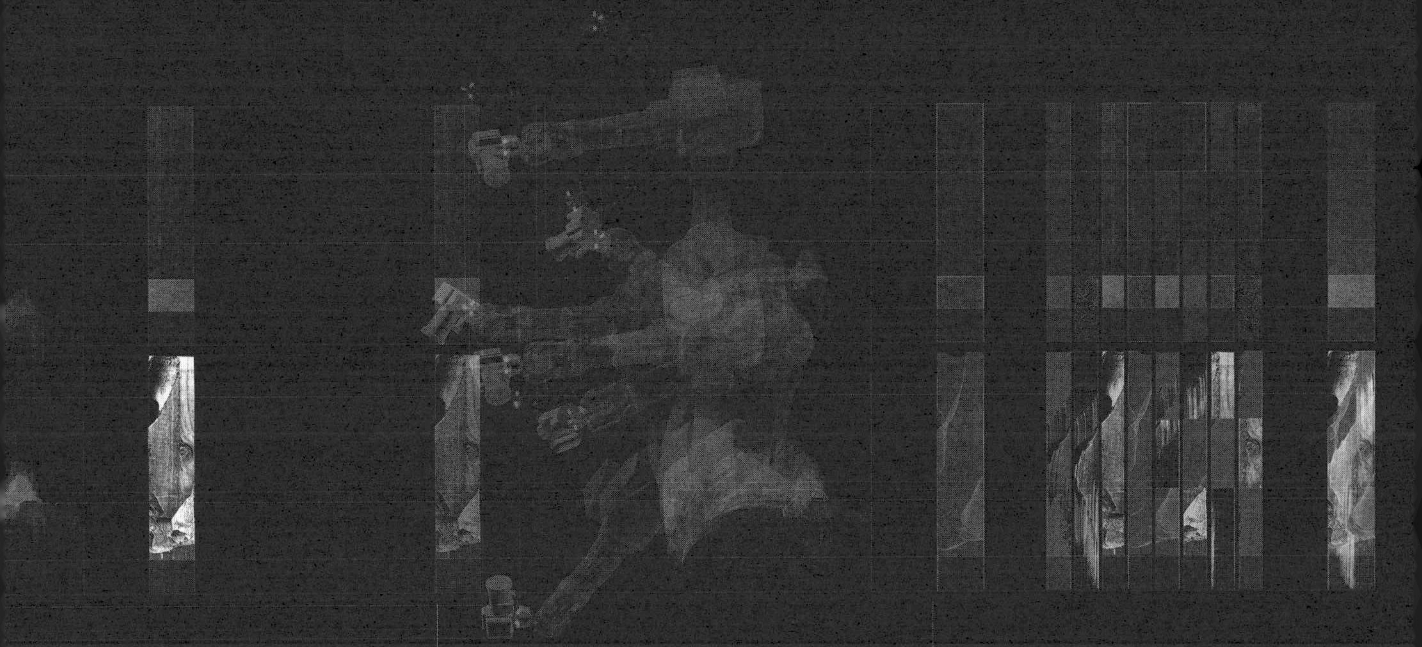
VERSE TRANSLATIONS
RANSLATIONS ASSEMBLY
ECONSTRUCTION

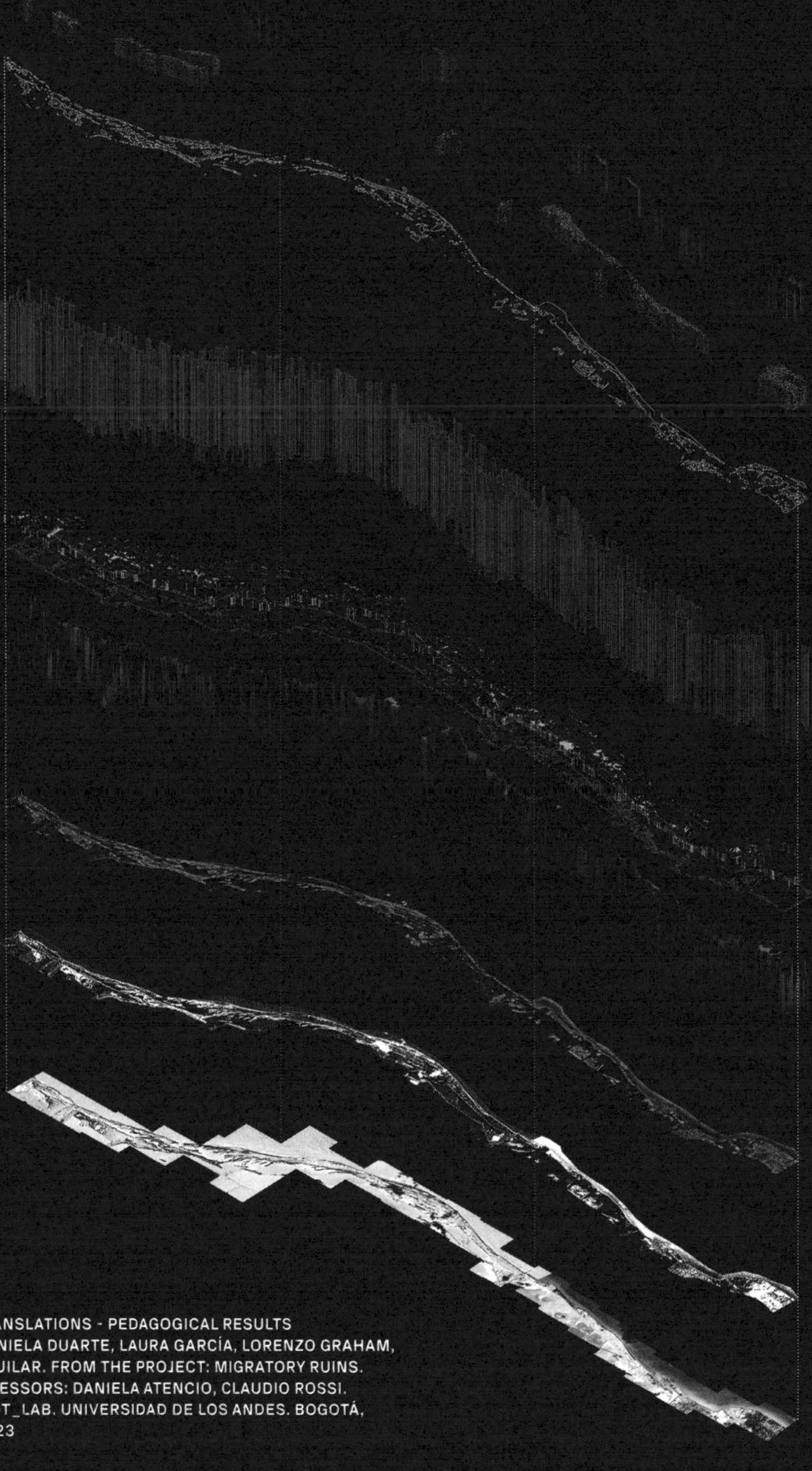

ANALOGUE TRANSLATIONS - PEDAGOGICAL RESULTS
STUDENTS: DANIELA DUARTE, LAURA GARCÍA, LORENZO GRAHAM, VALENTINA AGUILAR. FROM THE PROJECT: MIGRATORY RUINS. FACULTY-PROFESSORS: DANIELA ATENCIO, CLAUDIO ROSSI. COURSE: ROBOT_LAB. UNIVERSIDAD DE LOS ANDES. BOGOTÁ, COLOMBIA. 2023

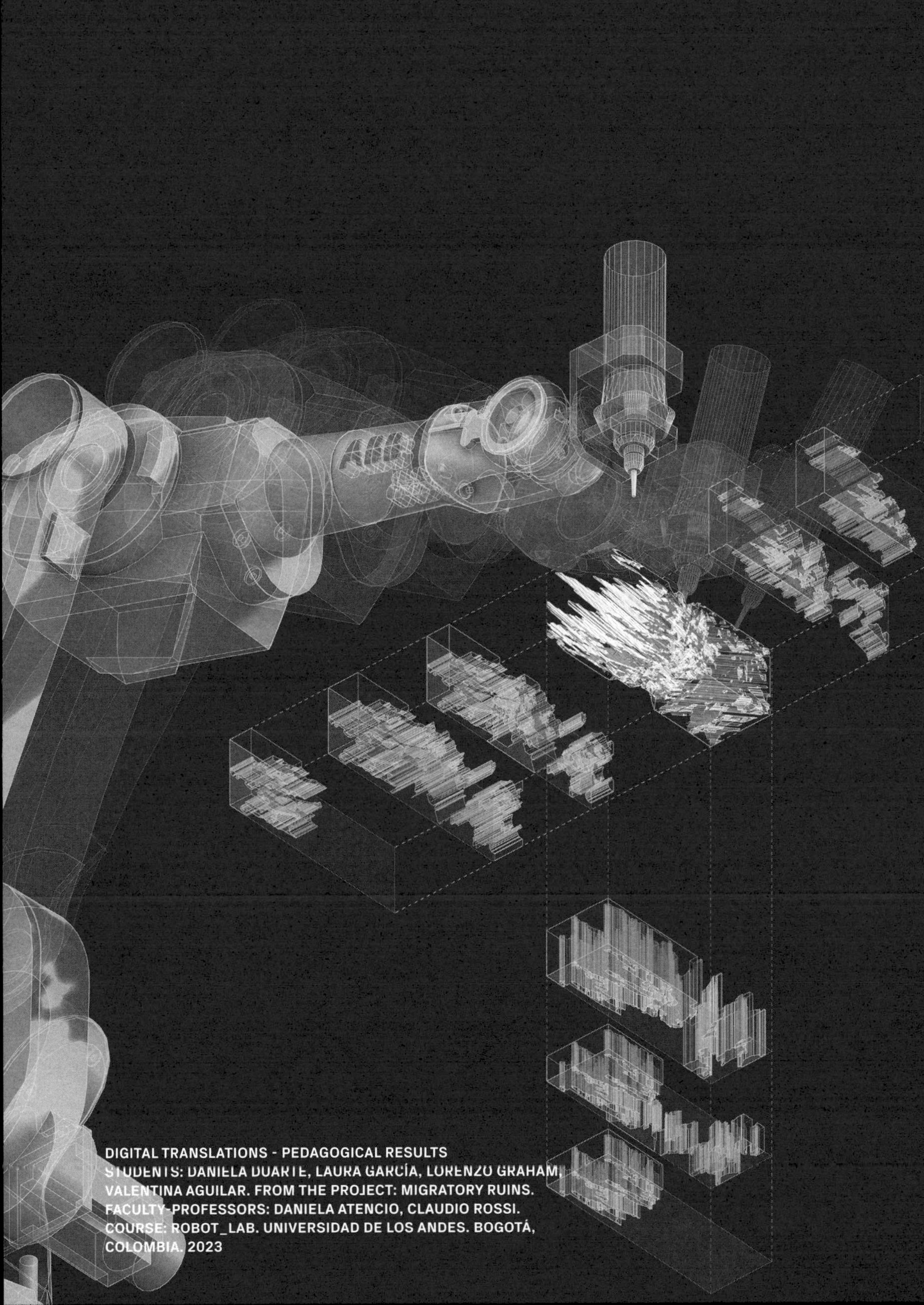

DIGITAL TRANSLATIONS - PEDAGOGICAL RESULTS
STUDENTS: DANIELA DUARTE, LAURA GARCÍA, LORENZO GRAHAM, VALENTINA AGUILAR. FROM THE PROJECT: MIGRATORY RUINS.
FACULTY-PROFESSORS: DANIELA ATENCIO, CLAUDIO ROSSI.
COURSE: ROBOT_LAB. UNIVERSIDAD DE LOS ANDES. BOGOTÁ, COLOMBIA. 2023

ROBOTIC [PHYSICAL] TRANSLATIONS - PEDAGOGICAL RESULTS
STUDENTS: DANIELA DUARTE, LAURA GARCÍA, LORENZO GRAHAM, VALENTINA AGUILAR. FROM THE PROJECT: MIGRATORY RUINS.
FACULTY-PROFESSORS: DANIELA ATENCIO, CLAUDIO ROSSI.
COURSE: ROBOT_LAB. UNIVERSIDAD DE LOS ANDES. BOGOTÁ, COLOMBIA. 2023

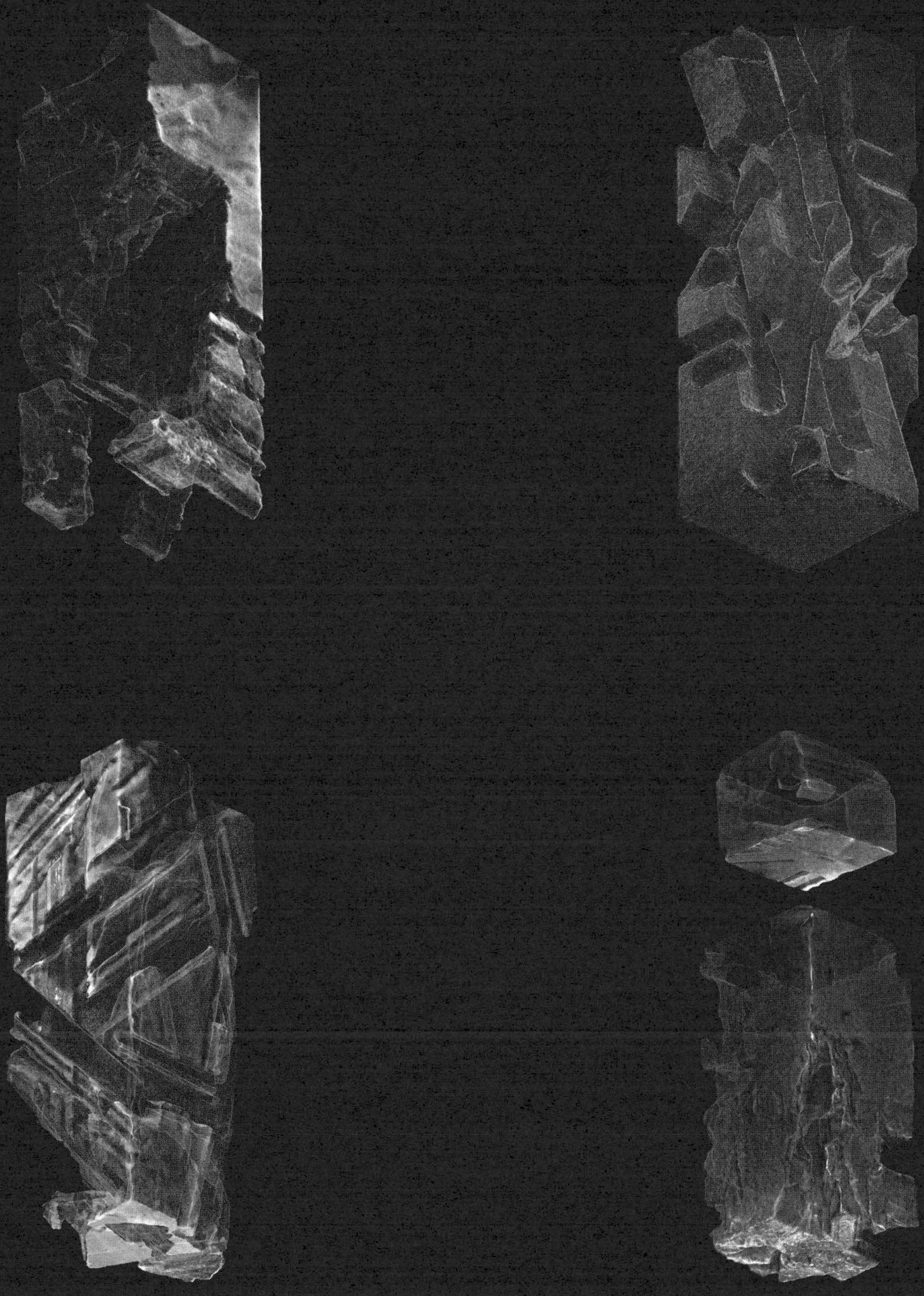

INVERSE [ROBOTIC] TRANSLATIONS - PEDAGOGICAL RESULTS STUDENTS: DANIELA DUARTE, LAURA GARCÍA, LORENZO GRAHAM, VALENTINA AGUILAR. FROM THE PROJECT: MIGRATORY RUINS. FACULTY-PROFESSORS: DANIELA ATENCIO, CLAUDIO ROSSI. COURSE: ROBOT_LAB. UNIVERSIDAD DE LOS ANDES. BOGOTÁ, COLOMBIA. 2023

TRANSLATIONS ASSEMBLIES: HYPER-ARTIFACT - PEDAGOGICAL RESULTS_CHUNK 1. STUDENTS: DANIELA DUARTE, LAURA GARCÍA, LORENZO GRAHAM, VALENTINA AGUILAR. FROM THE PROJECT: MIGRATORY RUINS. FACULTY-PROFESSORS: DANIELA ATENCIO, CLAUDIO ROSSI. COURSE: ROBOT_LAB. UNIVERSIDAD DE LOS ANDES. BOGOTÁ, COLOMBIA. 2023

TRANSLATIONS ASSEMBLIES: HYPER-ARTIFACT - PEDAGOGICAL RESULTS_CHUNK 2. STUDENTS: DANIELA DUARTE, LAURA GARCÍA, LORENZO GRAHAM, VALENTINA AGUILAR. FROM THE PROJECT: MIGRATORY RUINS. FACULTY-PROFESSORS: DANIELA ATENCIO, CLAUDIO ROSSI. COURSE: ROBOT_LAB. UNIVERSIDAD DE LOS ANDES. BOGOTÁ, COLOMBIA. 2023

ROBOTIC TRANSLATIONS_ LANDSCAPES CASE STUDY: MOJAVE DESERT

DANIELA ATENCIO
CLAUDIO ROSSI

The programmed landscape *Hyper-artifacts* were infrastructure in (and for) the Guajira Desert, Colombia; but also, as a speculative starting point, for the Mojave Desert, California, USA. The desert, as landscapes in motion, friction, erosion, and as a platform for resilient infrastructure for local communities to mitigate climate change, must be tested from multiple approaches or for different actors.

This is the kind of question from the field of research-creation that relies on technology. What will be the future of these landscapes in the next 20, 50, 100 years? How will the process of building artifacts that can be materialized from a robotic arm in distant desertic places be? Understanding local and temporal logic? Will it be possible to edit precise material capacities like *Watanabe*'s poem descriptions? In different places? With local observations? Before, we talked about salt and sea level rise, anthropic processes, and biological epicenters. We wanted to move these findings (and questions) from the Guajira desert to California.

To achieve that we traveled to the Mojave Desert in California and worked with the robotic arm[1] of the Southern California Institute of Architecture (SCI_Arc), Los Angeles, USA. We were prototyping and materializing hyper-artefacts promoting adaptations for the Californian desert context. It was an academic study-workshop trip and a week-long intensive robotic research experiment. We wanted to test different things and translate the workflow to see different results in a transversal reflection of two dissimilar places.

This space was defined as a transdisciplinary intensive studio that aimed to investigate, identify, and propose new approaches to the project from various disruptive patterns to promote a discussion on the future of this region. The technological emphasis has always impacted the processes and concepts that determine both design and aesthetic (and ethic) regimes in our field. In this case, testing a process of design, conceptually (translations) and systemically (workflow), was able to build differentiated bridges between the conditions of a place, its specificities, and the aesthetic result. In other words, there was also an aesthetic translation using the same working process tested in the Guajira. This, which would seem obvious from a generalist point of view, turns out to be interesting for our own conclusions since the exercise was generated for a projective reading from Latin America and not the other way around.

1 ABB® (*Asea Brown Boveri*) 6700, 300 Kg. Payload and 3,2 m reach. Over a 7m rail mount.

The starting point was the registration of a specific area of the Californian desert to understand geological conditions, sediments, types of nature, inhabitants (human and non-humans), topography, and material logics, among many other descriptions. Also, climate and chronological conditions of wind, temperature, or shadows. Using an advanced LiDAR scanner, the translation journey started.

In this case the groups of students (Robot-Lab[2] and the Master of Architectural Technologies) joined to discuss and apply the workflow presented in the very first day of the workshop. The robotic sessions —using the capacity of the 'giant robot' at SCI_Arc— ended up in multiple editions of large *MDF* blocks, translating the digital artifacts.

The *hyper-artifacts* tested the way in which the faces or surfaces of the edited object could recreate the erosion conditions of the desert or the possible ancestral memory (like 'scars') that determined these places from the water: *"Contrary to the archetype, a desert is not defined by the absence of water. The desert landscape is formed by the memory of water."*[3]

This space explored the importance of design research through the deployment of new tools, means and methodologies of working with technology, representation, and specifically, robotics as a design tool to build, transform and provoke new possibilities in architecture in our context (and other latitudes)

We have to understand ourselves as culturally digital and digitally cultural agents, which is why the reinterpretation and the disciplinary encounter with these scenarios open the reflection on the learning of our stylistic or conceptual issues from new perspectives (How would places be thought from new technologies?, From Artificial Intelligence?, How does a robot interpret and translate these historical/landscape logics?, What would be the way to build it?, among other considerations) to promote new inflections and representations. We cannot detach this technological reality from the most rooted issues of our discipline, but we must make the opening to new creative and culturally accessible dynamics, especially in our (and from) context.

This was an example of a translation, from south to north. Like a Worm-Eye View.

2 For this exercise Robot-Lab changed its name to Robot-LAX.

3 Desert X 23 curatorial statement by Diana Campbell. Coachella Valley, CA

DESERT Z WORKSHOP SITE. MOJAVE DESERT FOR LIDAR SCANNING.

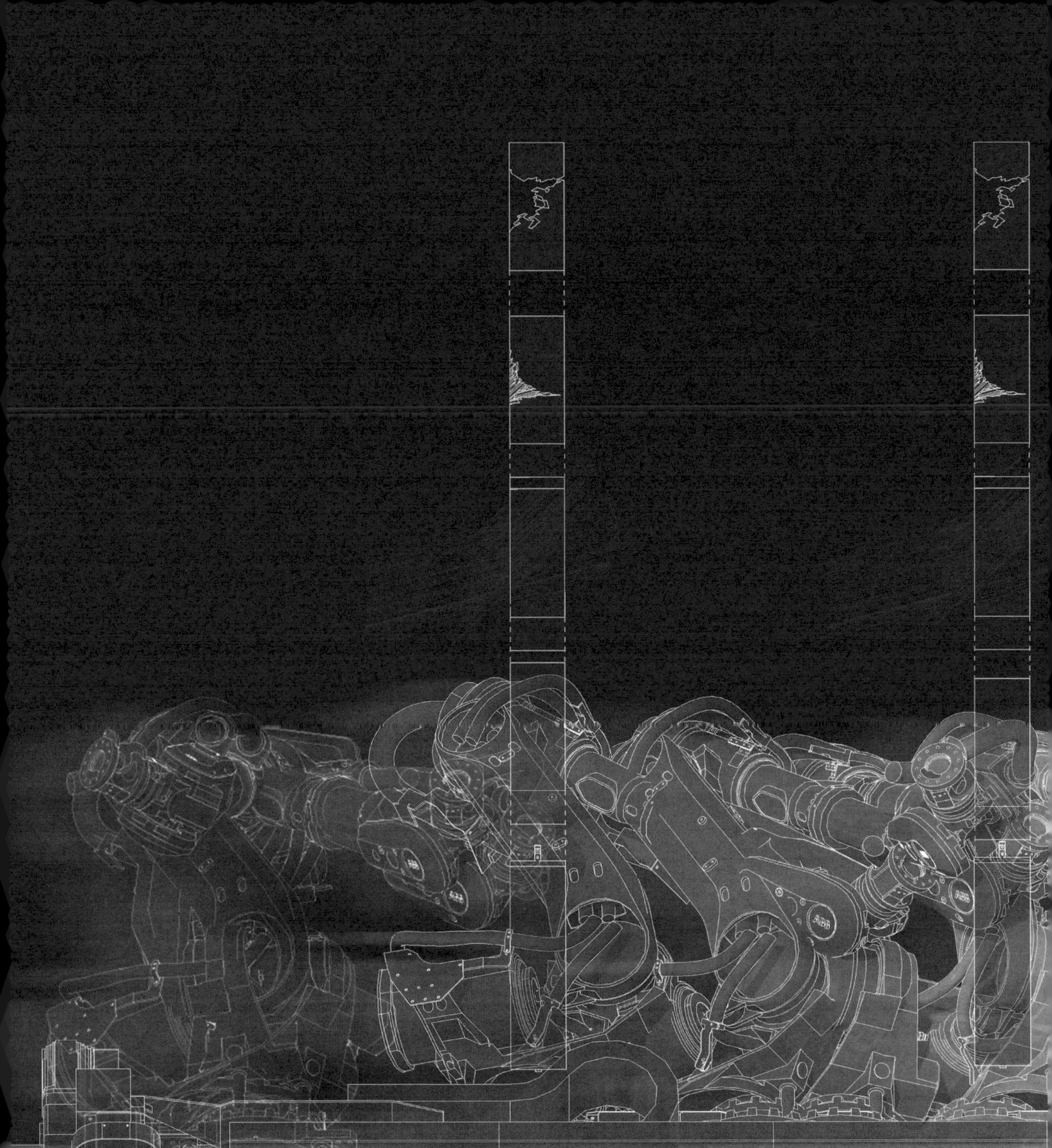

DESERT Z WORKSHOP SETUP. SCI_ARC ROBOT ARM ABB 6700 DRAWING IN WORM-EYE VIEW. CHOREOGRAPHY REPRESENTATION AND EXPLODED LAYERS OF EDITION. SIMULATION TOOLPATHS [FROM MOJAVE DESERT LIDAR SCANNING]

DESERT Z WORKSHOP SCENE. SCI_ARC ROBOT ARM ABB 6700. TOOL: MOTOTOOL. TOOLPATHS [FROM MOJAVE DESERT LIDAR SCANNING]

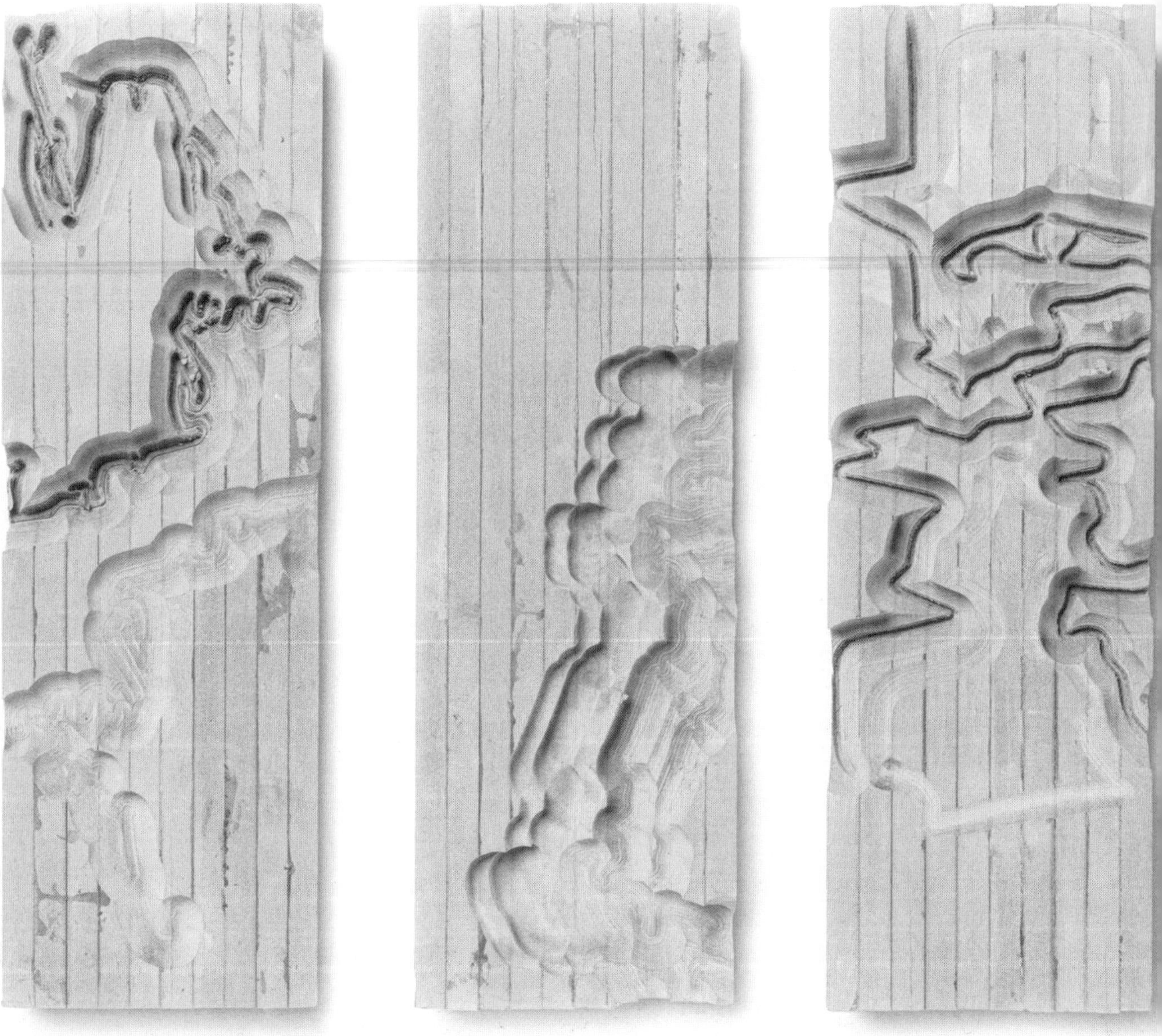

ROBOT_LAB: DESERT Z WORKSHOP - PEDAGOGICAL RESULTS_ ROBOTIC [PHYSICAL[TRANSLATIONS INDIVIDUAL PROTOTYPES.
STUDENTS: VALENTINA AGUAS, VALENTINA AGUILAR, LAURA ÁLVAREZ, ALEJANDRA DELGADO, DANIELA DUARTE, GABRIEL FIGUEROA, LAURA GARCÍA, SOFÍA GÓMEZ, LORENZO GRAHAM, SILVIA RUEDA, VALENTINA TRUJILLO, FELIPE ZORRO.
FACULTY-PROFESSORS: DANIELA ATENCIO, CLAUDIO ROSSI (UNIVERSIDAD DE LOS ANDES), CASEY REHM, SOOMEEN HAHM (SCI_ARC)
SOUTHERN CALIFORNIA INSTITUTE OF ARCHITECTURE + UNIVERSIDAD DE LOS ANDES. 2023

ROBOT_LAB: DESERT Z WORKSHOP - PEDAGOGICAL RESULTS_ ROBOTIC [PHYSICAL[TRANSLATIONS PROTOTYPES ASSEMBLIES [1].
STUDENTS: VALENTINA AGUAS, VALENTINA AGUILAR, LAURA ÁLVAREZ, ALEJANDRA DELGADO, DANIELA DUARTE, GABRIEL FIGUEROA, LAURA GARCÍA, SOFÍA GÓMEZ, LORENZO GRAHAM, SILVIA RUEDA, VALENTINA TRUJILLO, FELIPE ZORRO.
FACULTY-PROFESSORS: DANIELA ATENCIO, CLAUDIO ROSSI (UNIVERSIDAD DE LOS ANDES), CASEY REHM, SOOMEEN HAHM (SCI_ARC)
SOUTHERN CALIFORNIA INSTITUTE OF ARCHITECTURE + UNIVERSIDAD DE LOS ANDES. 2023

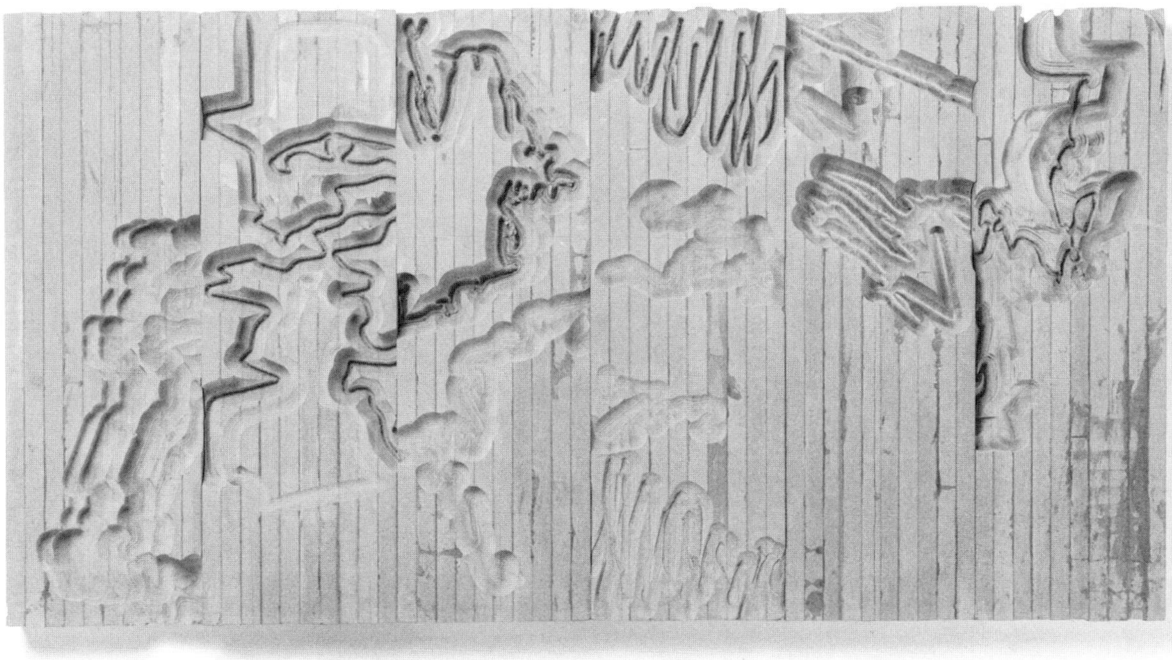

ROBOT_LAB: DESERT Z WORKSHOP - PEDAGOGICAL RESULTS_ ROBOTIC [PHYSICAL[TRANSLATIONS PROTOTYPES ASSEMBLIES [2].
STUDENTS: VALENTINA AGUAS, VALENTINA AGUILAR, LAURA ÁLVAREZ, ALEJANDRA DELGADO, DANIELA DUARTE, GABRIEL FIGUEROA, LAURA GARCÍA, SOFÍA GÓMEZ, LORENZO GRAHAM, SILVIA RUEDA, VALENTINA TRUJILLO, FELIPE ZORRO.
FACULTY-PROFESSORS: DANIELA ATENCIO, CLAUDIO ROSSI (UNIVERSIDAD DE LOS ANDES), CASEY REHM, SOOMEEN HAHM (SCI_ARC)
SOUTHERN CALIFORNIA INSTITUTE OF ARCHITECTURE + UNIVERSIDAD DE LOS ANDES. 2023

ROBOT_LAB: DESERT Z WORKSHOP - PEDAGOGICAL RESULTS_PROTOTYPE CLOSE-UP [1]. STUDENTS: SILVIA RUEDA. FACULTY-PROFESSORS: DANIELA ATENCIO, CLAUDIO ROSSI (UNIVERSIDAD DE LOS ANDES), CASEY REHM, SOOMEEN HAHM (SCI_ARC). SOUTHERN CALIFORNIA INSTITUTE OF ARCHITECTURE + UNIVERSIDAD DE LOS ANDES. 2023

ROBOT_LAB: DESERT Z WORKSHOP - PEDAGOGICAL RESULTS_PROTOTYPE CLOSE-UP [2].STUDENTS: SILVIA RUEDA. FACULTY-PROFESSORS: DANIELA ATENCIO, CLAUDIO ROSSI (UNIVERSIDAD DE LOS ANDES), CASEY REHM, SOOMEEN HAHM (SCI_ARC). SOUTHERN CALIFORNIA INSTITUTE OF ARCHITECTURE + UNIVERSIDAD DE LOS ANDES. 2023

ROBOTIC CIRCUMSTANCES: DESERT Z EXHIBITION. MATERIAL CLOSE-UP DURING EXHIBITION. 2023

ROBOTIC TOOL-KIT: MACHINES FOR MACHINES

OR [THE TRANSLATORS]

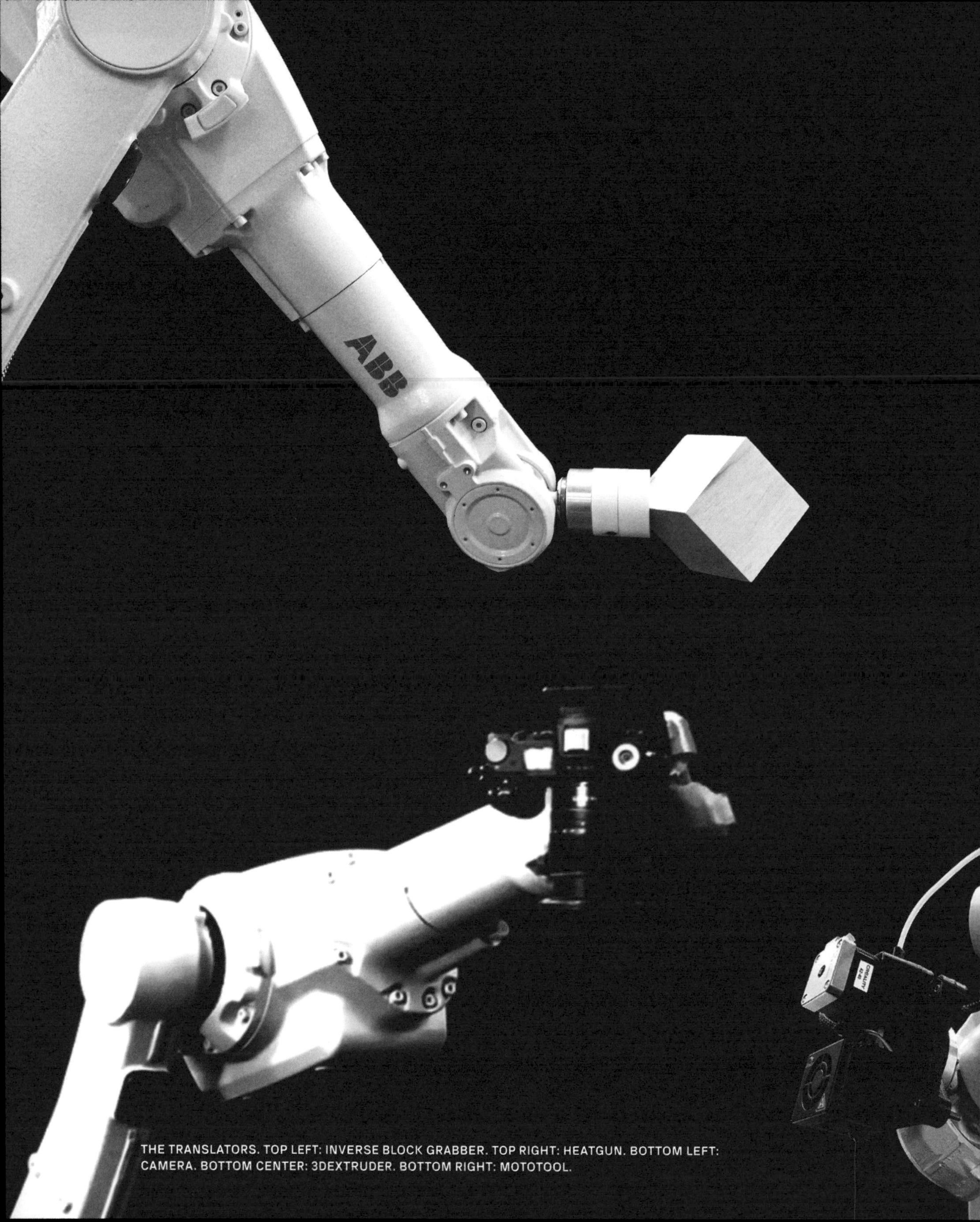

THE TRANSLATORS. TOP LEFT: INVERSE BLOCK GRABBER. TOP RIGHT: HEATGUN. BOTTOM LEFT: CAMERA. BOTTOM CENTER: 3DEXTRUDER. BOTTOM RIGHT: MOTOTOOL.

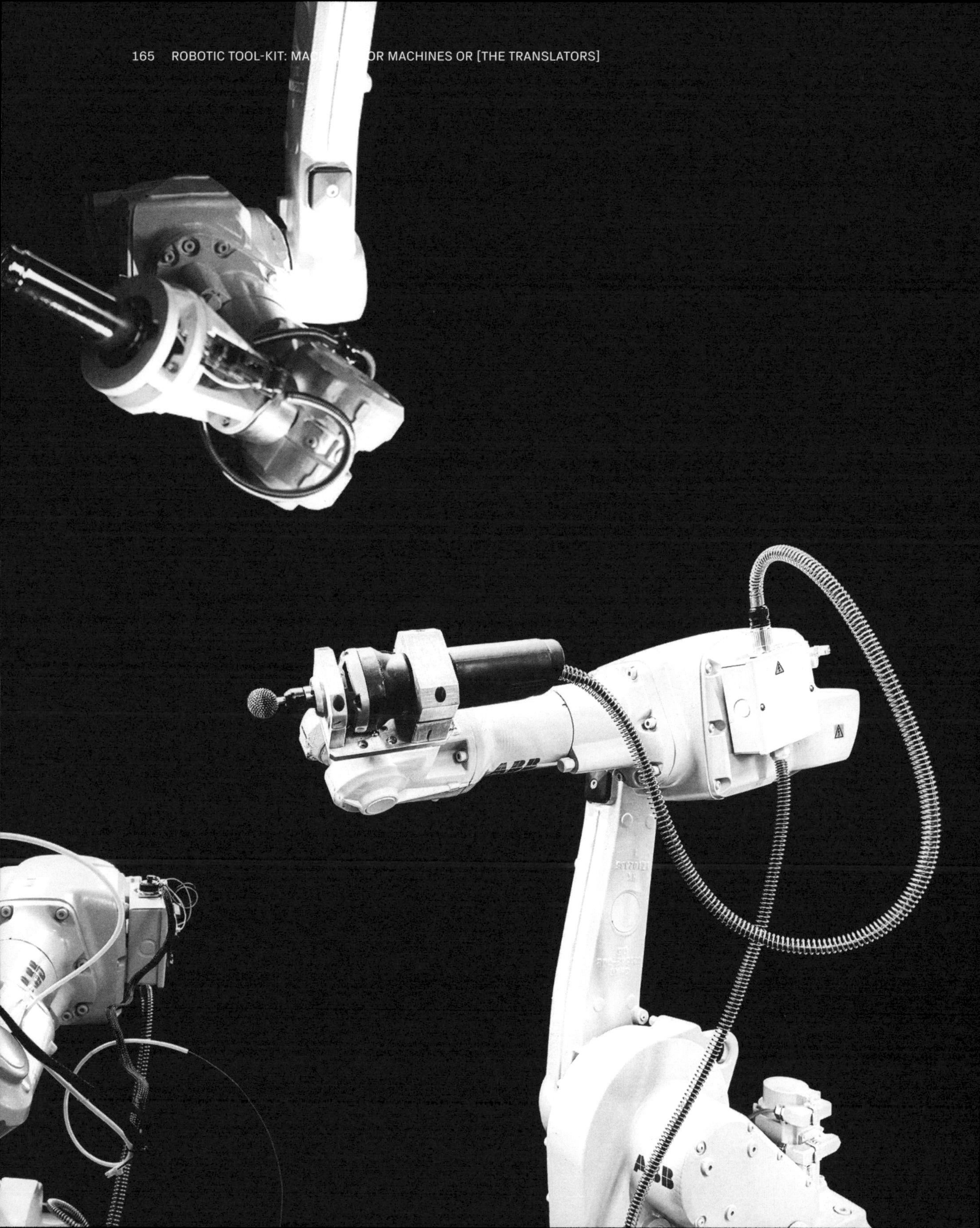

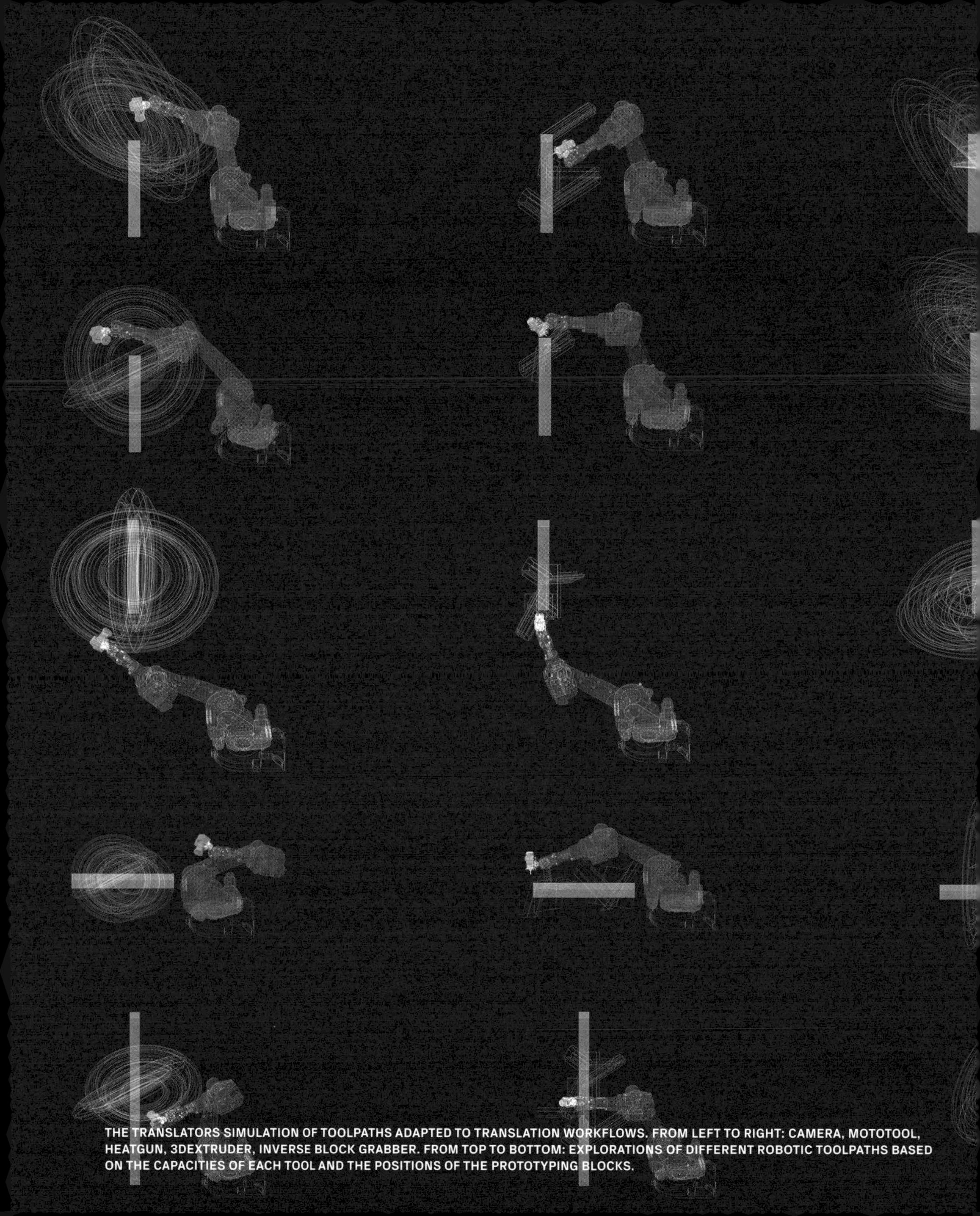

THE TRANSLATORS SIMULATION OF TOOLPATHS ADAPTED TO TRANSLATION WORKFLOWS. FROM LEFT TO RIGHT: CAMERA, MOTOTOOL, HEATGUN, 3DEXTRUDER, INVERSE BLOCK GRABBER. FROM TOP TO BOTTOM: EXPLORATIONS OF DIFFERENT ROBOTIC TOOLPATHS BASED ON THE CAPACITIES OF EACH TOOL AND THE POSITIONS OF THE PROTOTYPING BLOCKS.

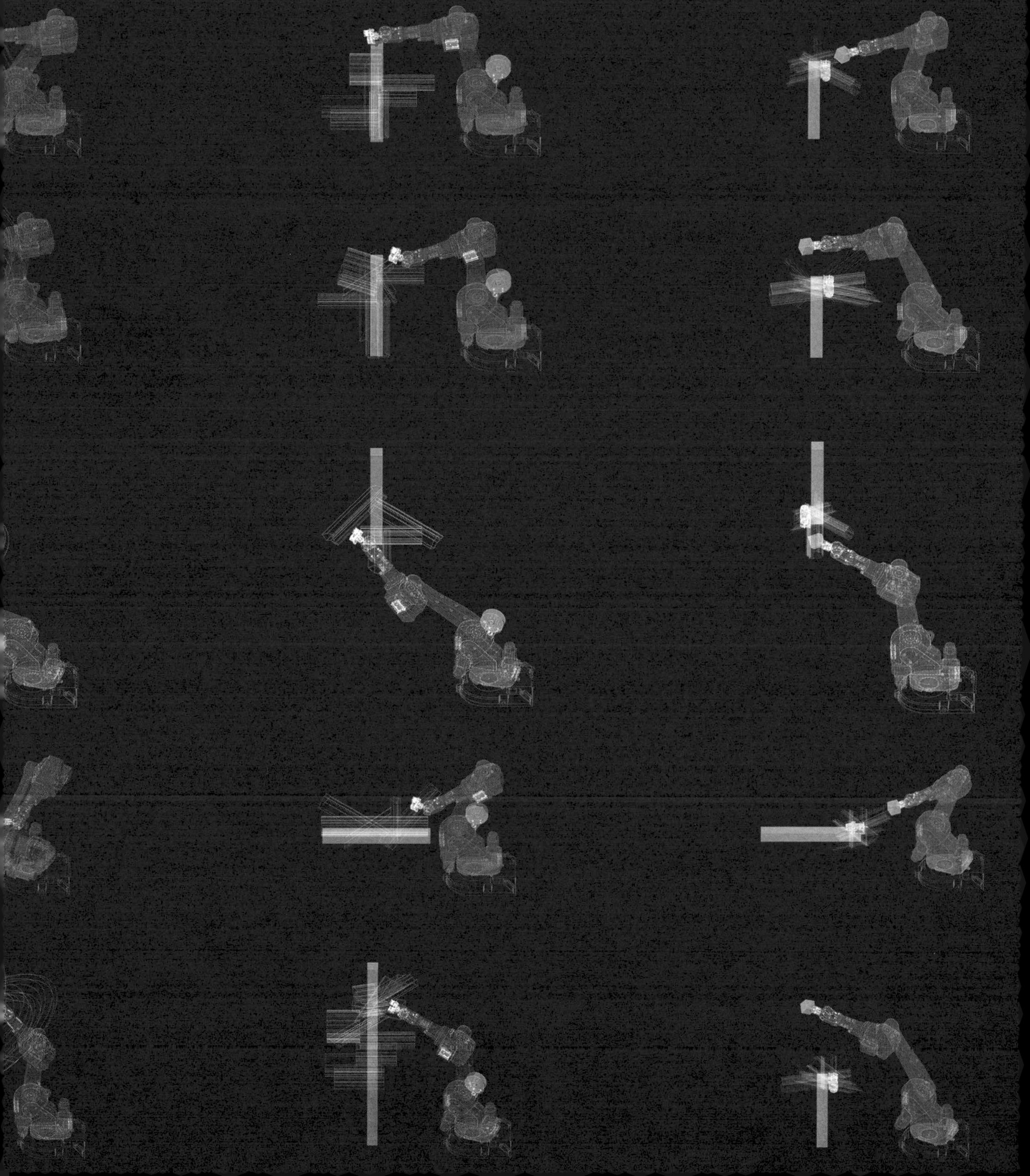

TRANSLATOR 001_MOTOTOOL

Model: Bosch® GGS 28 CE / 30.000 RPM

Capacities: This tool was used to edit the MDF blocks, performing the toolpaths according to the workflow (See translations). Working with this machine allowed the robot to test different bits, some of which were reported in this tool kit. The other interesting possibility is the use of different RPMs to see the level of edition in the MDF block.

Actions: Edit, Burr, Grave, Carve, Mill, Grind, Transform, Modify, Erode, Devastate, Corrode, Mold, Extract, Texturize, Subtract, Engrave, Buff, Drill, Ruin, Gnaw .

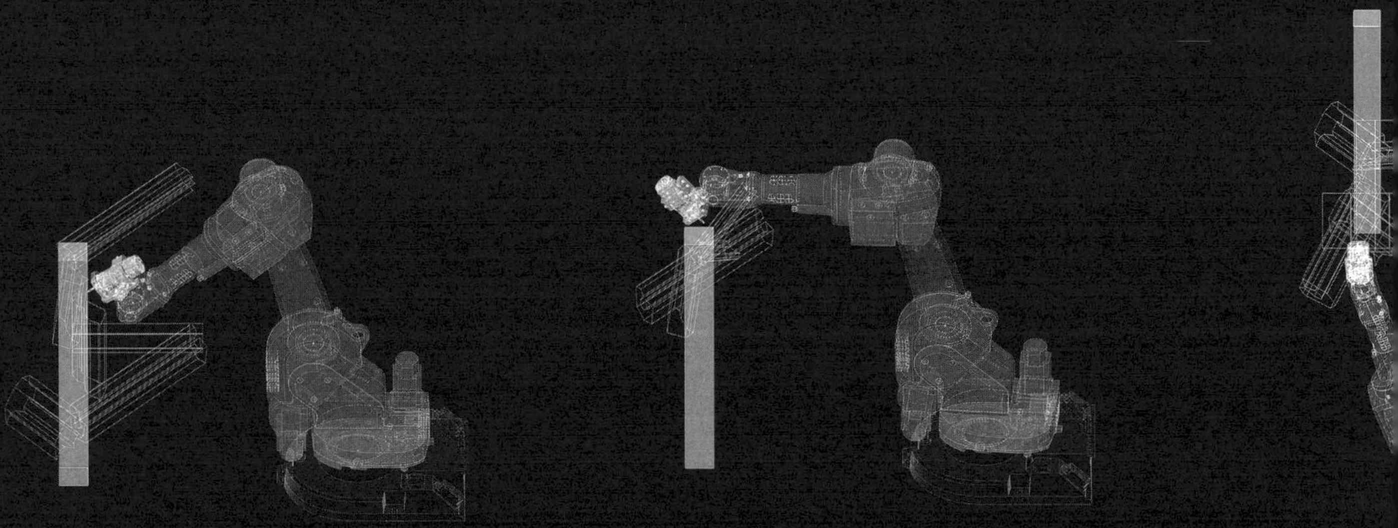

THE TRANSLATOR 001 [MOTOTOOL]. SIMULATION OF TOOLPATHS ADAPTED TO TRANSLATION WORKWORKFLOWS.

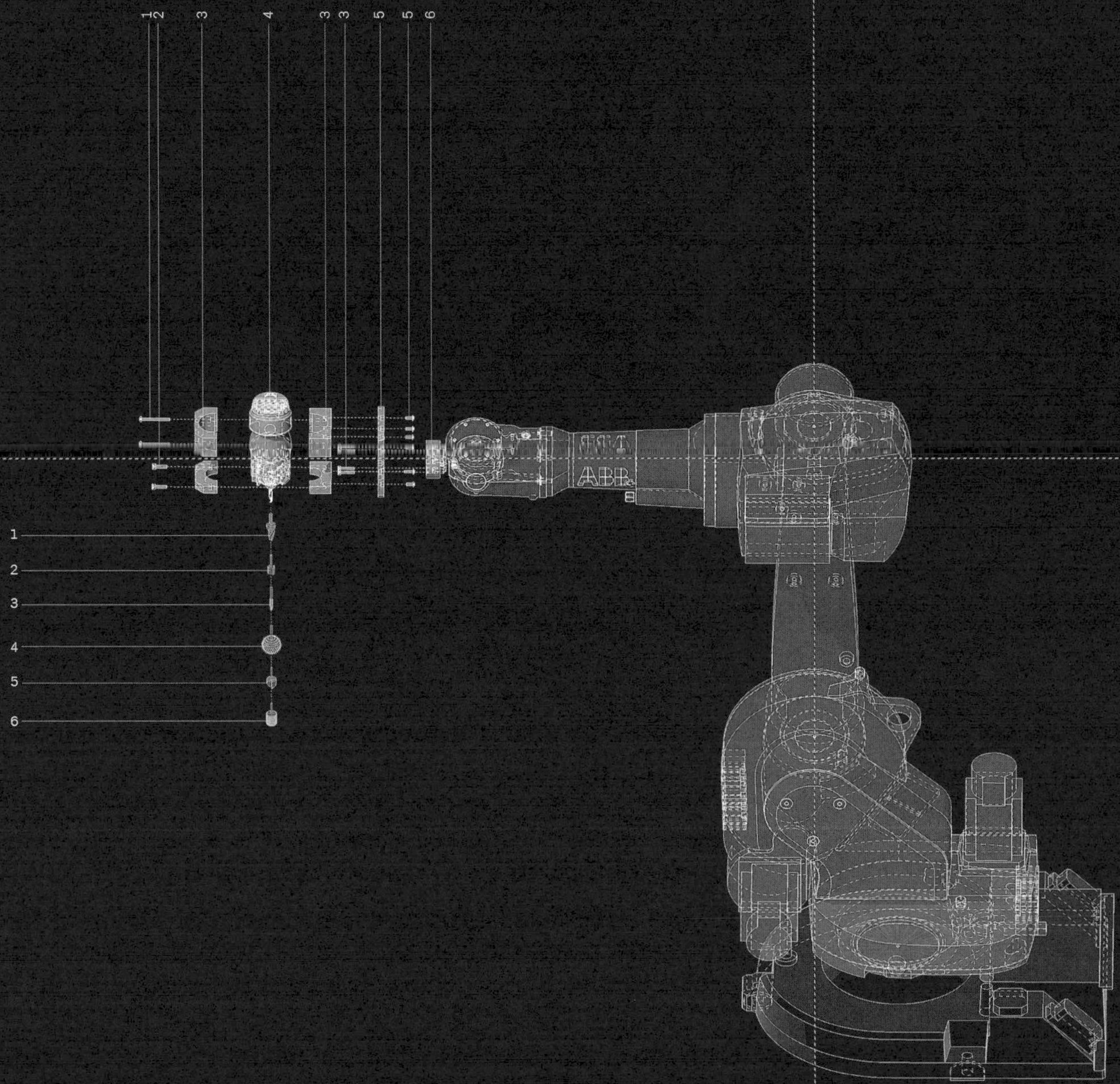

Assembly pieces

1 2 X M3 X 25 Screws [2]
2 M3 X 16 Screws 2
3 ½" Aluminum Brackets, 2 X M6 X 25 Screws [2]
4 Bosch® GGS 28 CE / 30.000 RPM
5 ½" Aluminum Plate, 4 X M5 X 20 Screws [6]
6 1" Aluminum Spacer, D: Ø 2"

LEFT: ABB 1600 ROBOT + TRANSLATOR 001: MOTOTOOL. OBLIQUE DRAWING WITH EXPLODED PARTS. RIGHT: ABB 1600 ROBOT + TRANSLATOR 001: MOTOTOOL. OBLIQUE DRAWING + WORM-EYE VIEW DRAWING WITH EXPLODED PARTS.

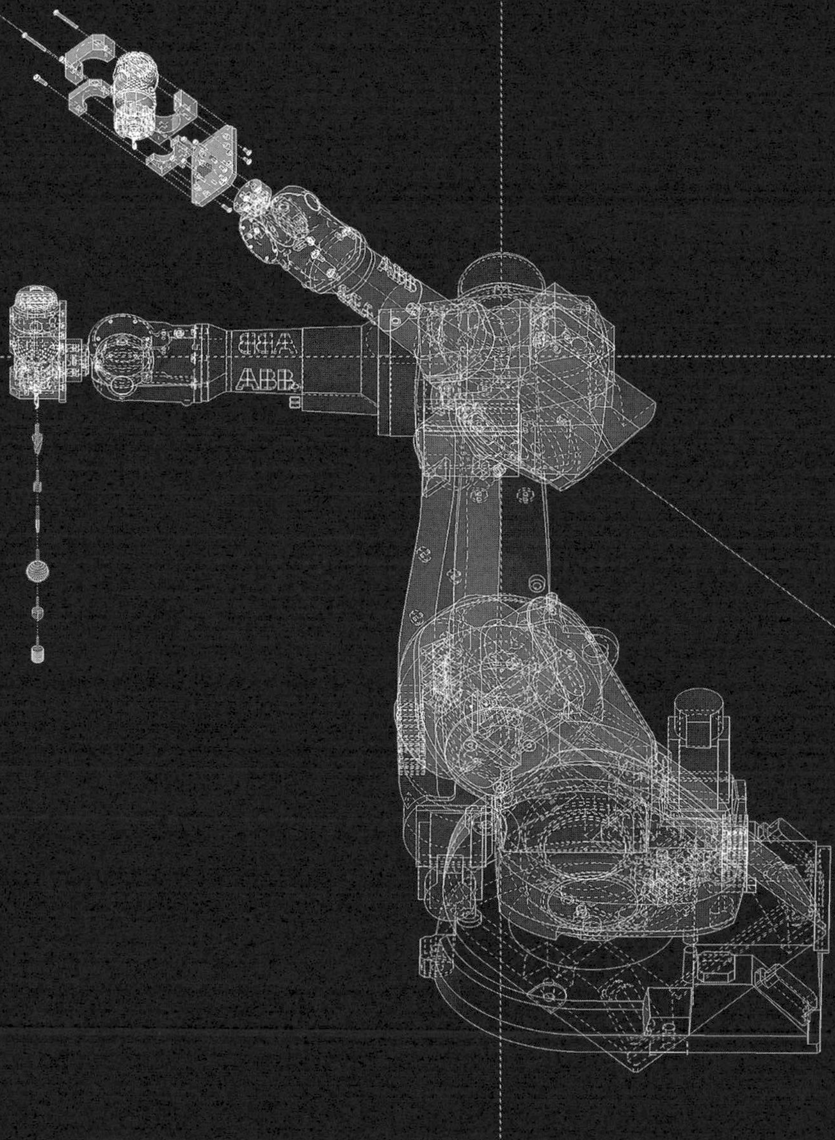

Drill/Burr Bits

1 Solid Carbide V Groove Engraving 60o, Ø ¼" Shaft, D: Ø 0.025"
2 Shank 2-Flute Carbide Tipped, Ø ¼" Shaft, D: ¾" (Ø 19mm)
3 Taper Rotary Burr, Ø ¼" Shaft, Abrasive Tungsten Carbide, D: ¼" (Ø 6.3mm)
4 Sphere Rotary Burr, Ø ¼" Shaft, Abrasive Tungsten Carbide, D: 1" (Ø 25.4mm)
5 Ball Nose Rotary Burr, Ø ¼" Shaft, Abrasive Tungsten Carbide, D: ¾" (Ø 19mm)
6 Cylinder Rotary Burr, Ø ¼" Shaft, Abrasive Tungsten Carbide, D: ¾" (Ø 19mm)
[Among others]

TRANSLATOR 002_CAMERA

Model: Nikkon® Z 14-30mm f/4 S

Capacities: This translator was used and made to achieve the robotic choreography: a continuous movement with the camera and the robot, capturing the newly generated geometry. The camera record generated a point cloud and meshed 3D model. We did it by m°delling a support for the Nikkon® camara in PLA to make possible the free movements of the toolpaths in a safe way, —considering the weight of the camera and the lenses.

Actions: Move, Choreograph, Flow, Transport, Record, Capture, Document, Register, Reinterpret, Scan, Analyze, Track, Follow.

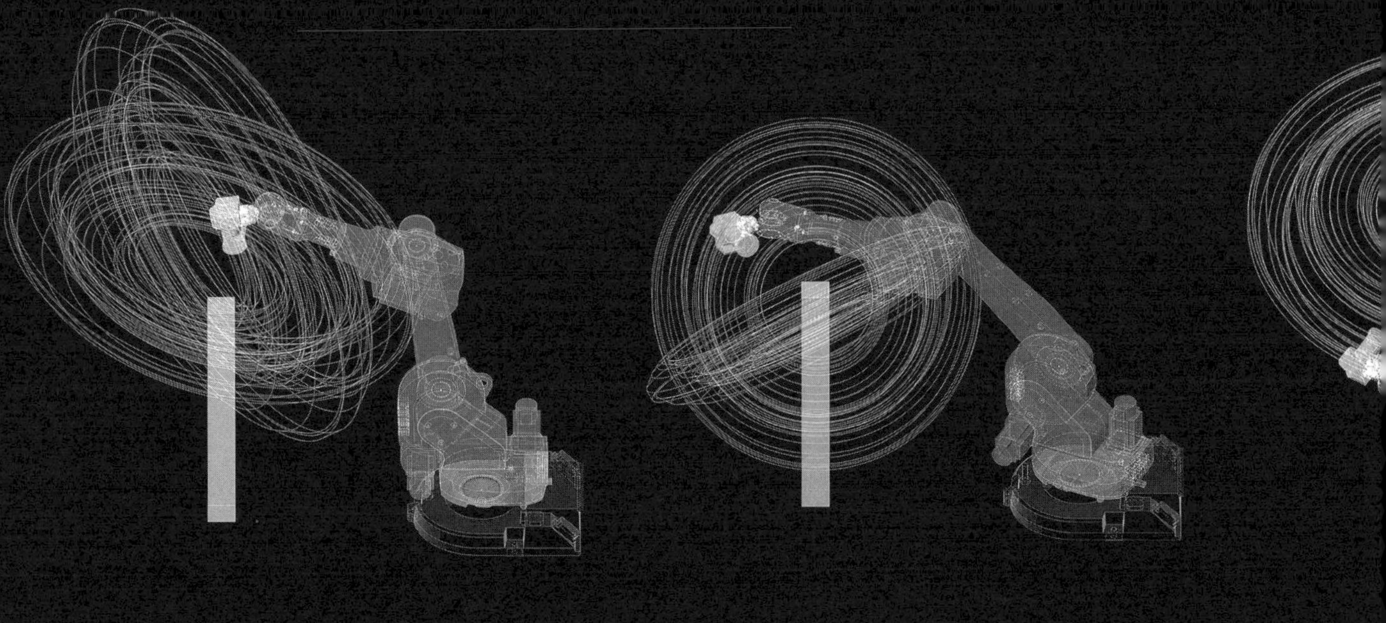

THE TRANSLATOR 002 [CAMERA]. SIMULATION OF TOOLPATHS ADAPTED TO TRANSLATION WORKFLOWS.

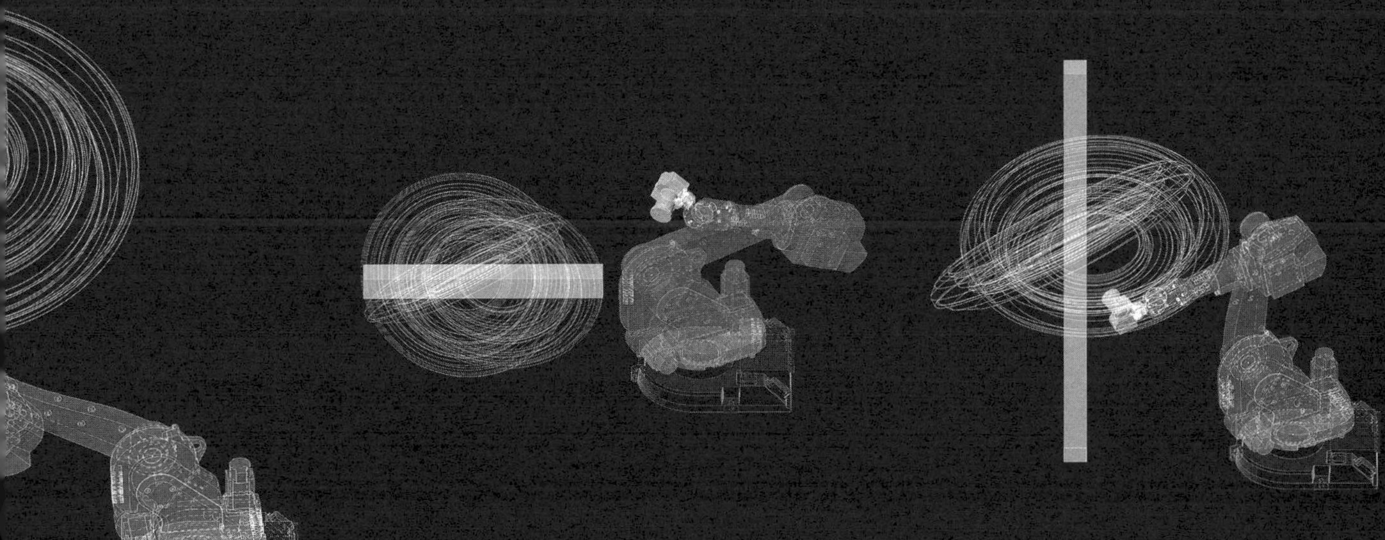

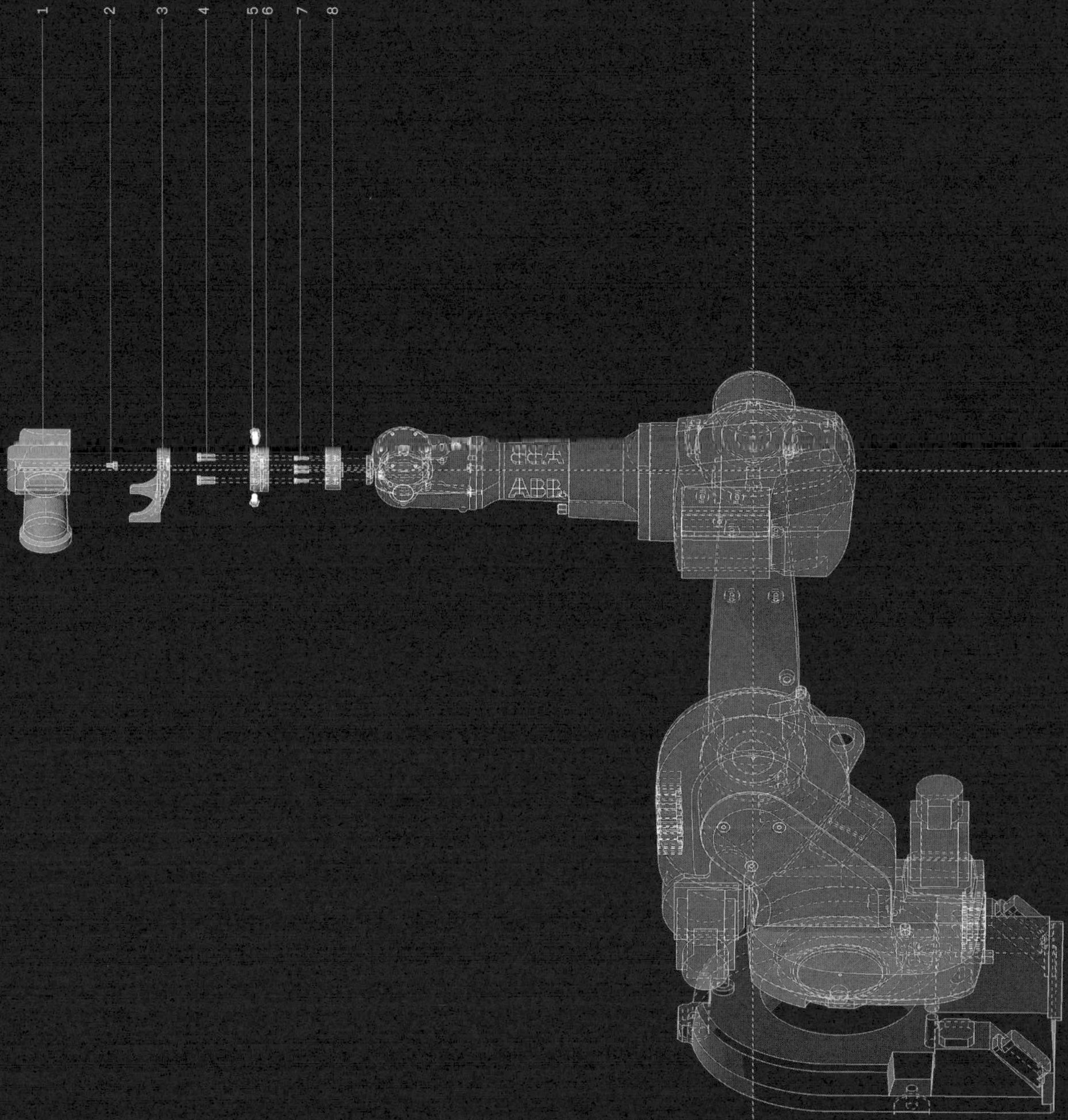

Assembly pieces

1 *Nikkon® Z 14-30mm f/4 S*
2 Camera Screw Ø ¼"
3 Custom designed 3DPrinted Camera Holder [PLA]
4 2 X M3 X 25 Screws [4]
5 Screws Ø ½" [3]
6 Custom designed 3DPrinted Tool Plate [PLA]
7 2 X M6 X 25 Screws [4]
8 1" Aluminum Spacer, D: Ø 2"

LEFT: ABB 1600 ROBOT + TRANSLATOR 002:CAMERA. OBLIQUE DRAWING WITH EXPLODED PARTS. RIGHT: ABB 1600 ROBOT + TRANSLATOR 002: CAMERA. OBLIQUE DRAWING + WORM-EYE VIEW DRAWING WITH EXPLODED PARTS.

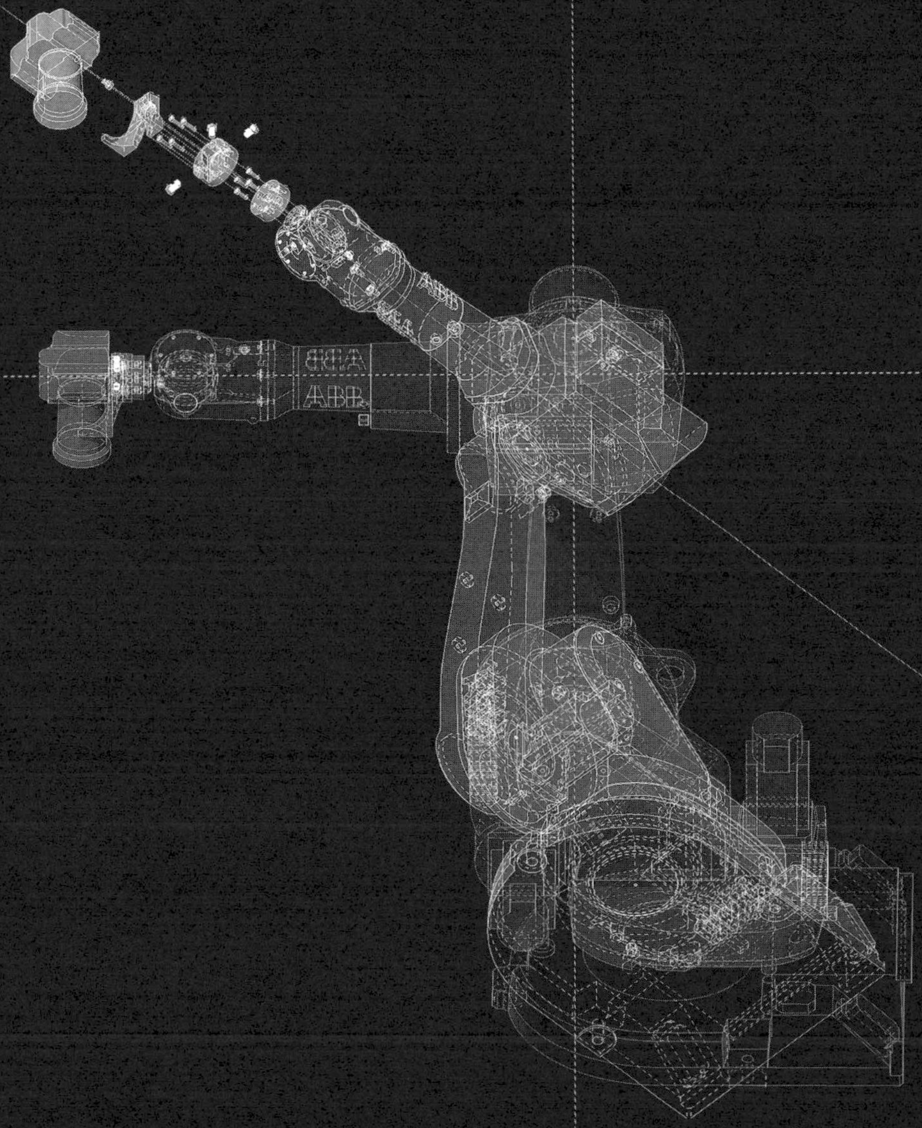
ABB

TRANSLATOR 003_HEATGUN

Model: Custom made

Capacities: We use the heat resistance and the blowing mechanism of the heavy-duty heat gun's heat resistance and blowing mechanism. From this, several tests were made to understand temperatures, watts/amps, and the PLA resistance since the heat gun needed to be running for several hours. Finally, a two-part PLA custom design attachment for the robot was made to support the heat gun nozzle and the board that controlled temperature, blowing speed, and the on/off switch. The parts were attached together with a wide-open system that allowed the nozzle not to overheat or melt the tool. The other specific design for this tool was the set of wires that needed to be as free as the movements involved in the toolpath, without touching the heat, but also that moved around freely without struggling with joint movements.

Actions: Melt, Evaporate, Heat, Fade, Soften, Warm, Diffuse, Fuse, Convert, Alternate, Thaw, Fuse, Deform, Vanish, Liquefy, Remove, Warp

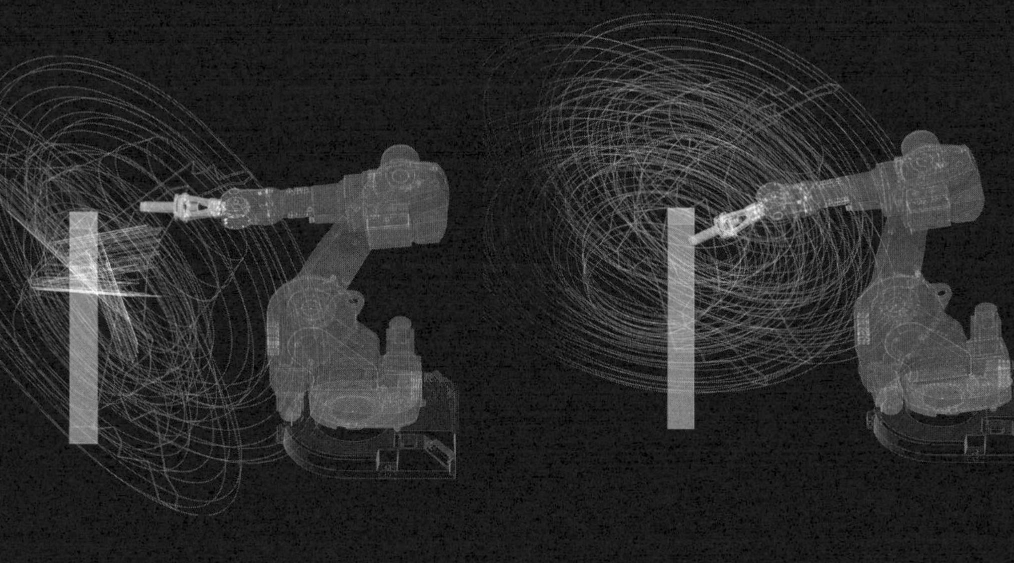

THE TRANSLATOR 003 [HEATGUN]. SIMULATION OF TOOLPATHS ADAPTED TO TRANSLATION WORKFLOWS.

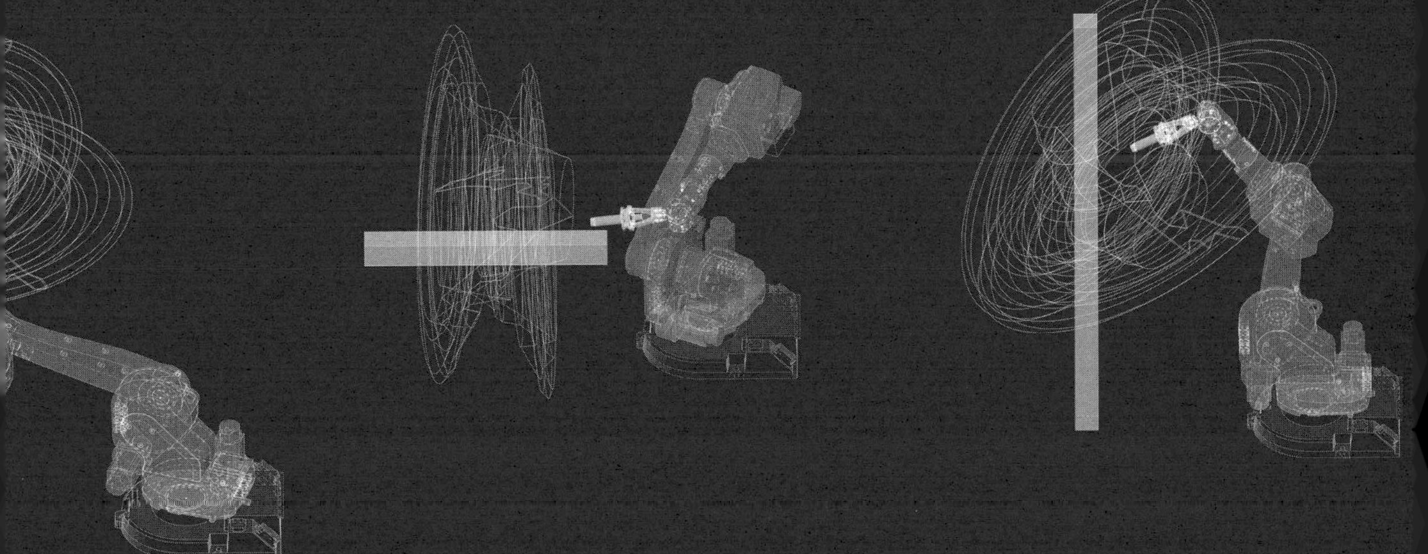

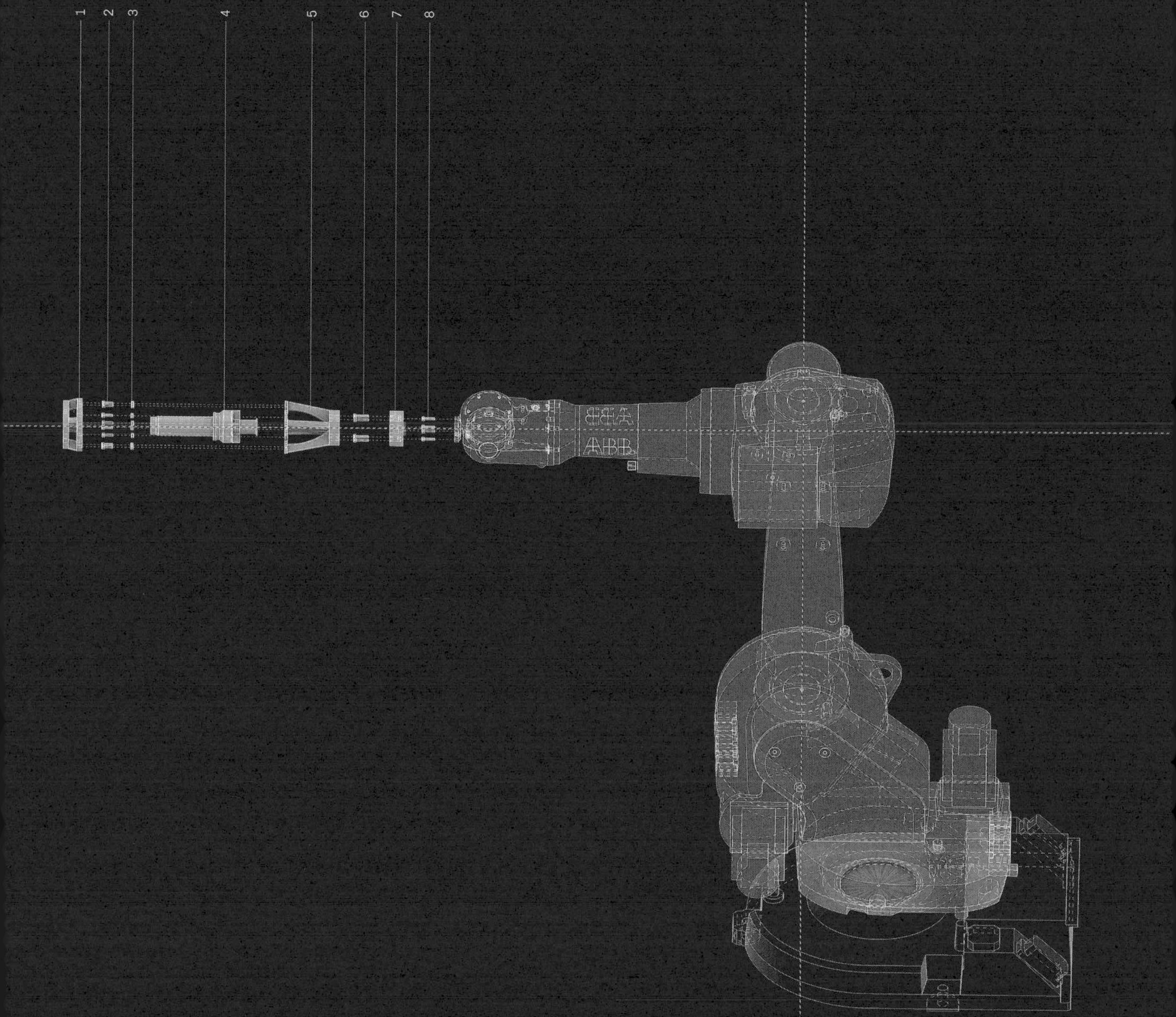

Assembly pieces
Part A
1 Custom designed 3DPrinted Heater Holder [PLA]
2 4 X M5 X 16 Screws [6]
3 4 X M5 Hex Nut [6]
4 Heat gun nozzle

LEFT: ABB 1600 ROBOT + TRANSLATOR 003: HEATGUN. OBLIQUE DRAWING WITH EXPLODED PARTS. RIGHT: ABB 1600 ROBOT + TRANSLATOR 003: HEATGUN. OBLIQUE DRAWING + WORM-EYE VIEW DRAWING WITH EXPLODED PARTS.

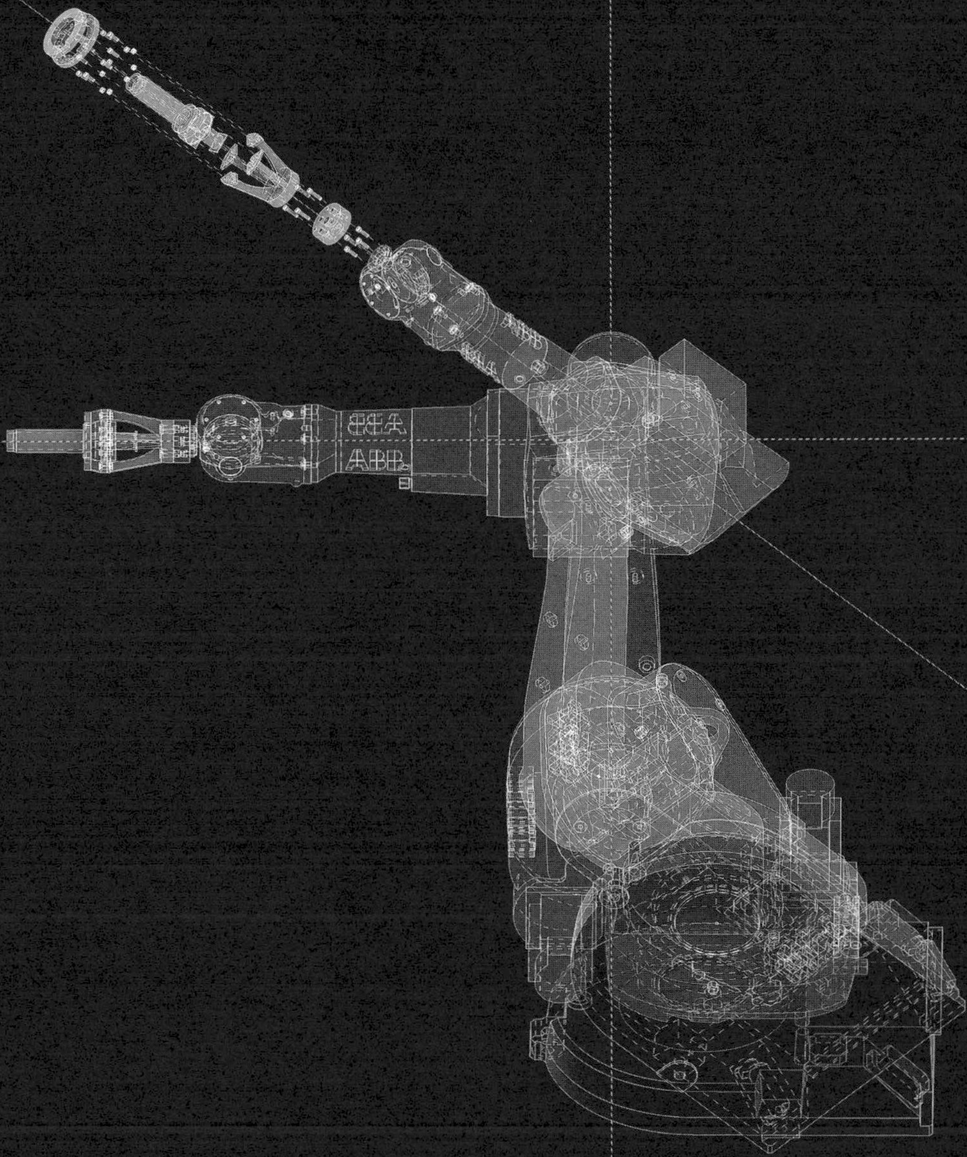

Part B
5 Custom designed 3DPrinted Tool Plate [PLA]
6 4 X M5 X 20 Screws [4]
7 1" Aluminum Spacer, D: Ø 2"
8 4 X M5 X 20 Screws [4]

TRANSLATOR 004_3DEXTRUDER

Model: Custom made.

Capacities: A custom-made 3DExtruder assembly was made to test a different kind of translation. In this case, and using the actions (verbs) proposed, the idea was to accumulate PLA geometries via material deposition to understand the kind of fabrication and possible robotic physical translations. The tests provided a path to challenge different kinds of sustainable/recycled plastics at different scales.

Actions: Overlay, Overlap, Superimpose, Merge, Fuse, Attach, add, Unite, Melt, Blend, Push, Compress, Slice, Agglomerate, Accumulatep

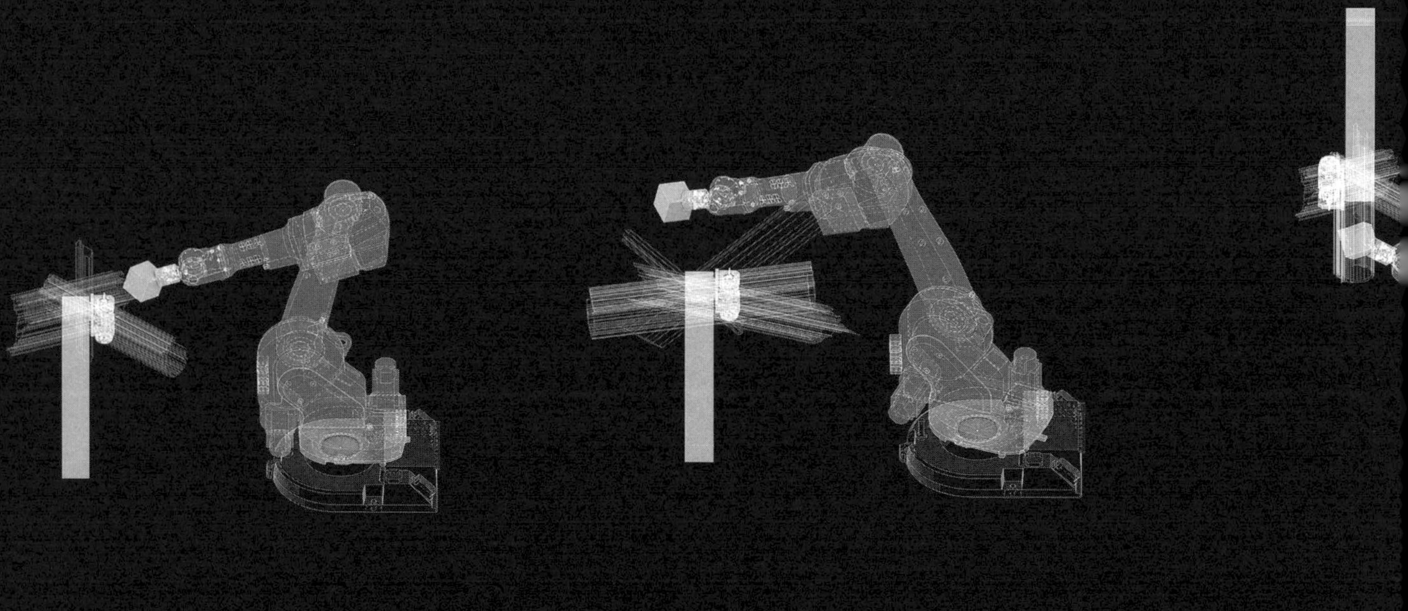

THE TRANSLATOR 004 [3DEXTRUDER]. SIMULATION OF TOOLPATHS ADAPTED TO TRANSLATION WORKFLOWS.

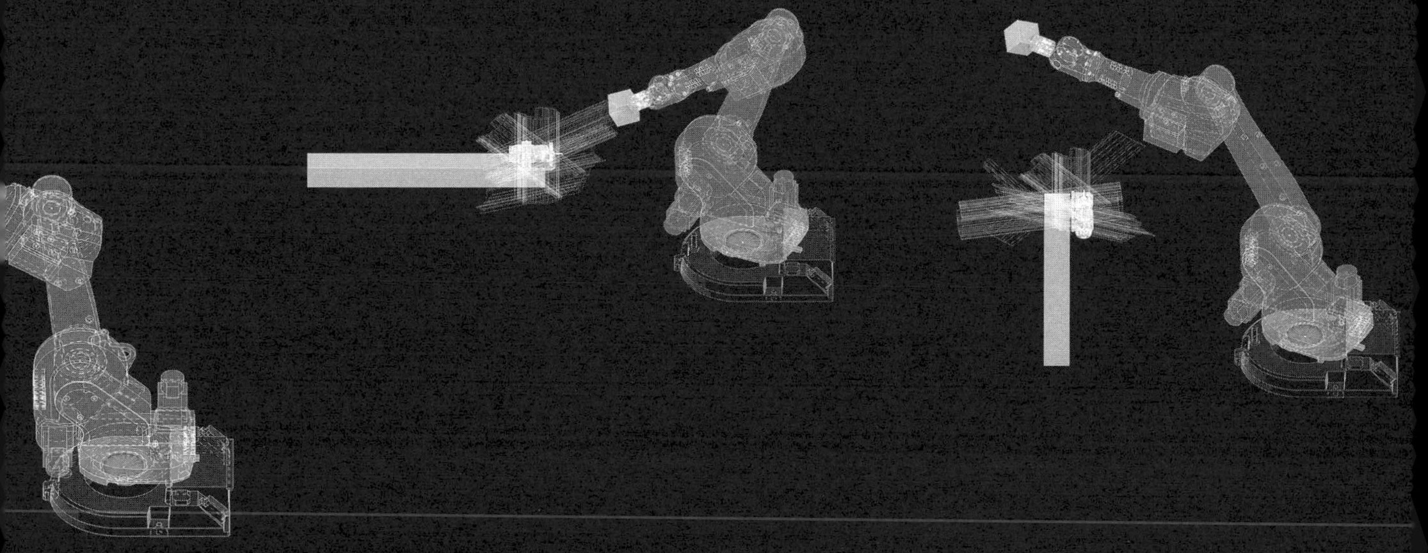

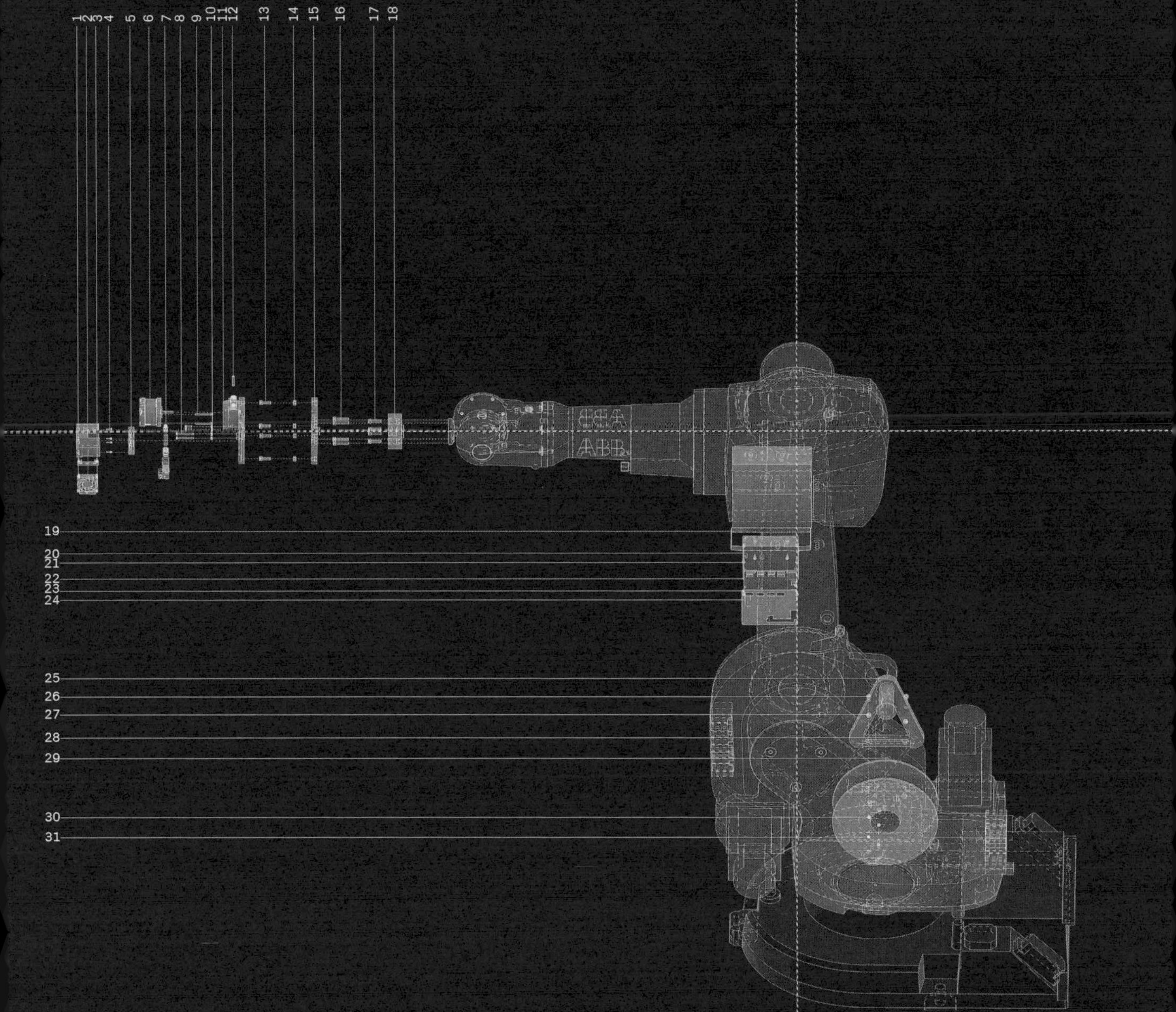

Assembly pieces

Extruder

1 Fan 40 X 40mm
2 Fan case
3 Thermistor support
4 4 X M5 X 16
5 Custom designed 3DPrinted Cap (PLA)
6 Nema 40/20 Stepper
7 MKB Hotend
8 4 X M5 X 20
9 2 X M3 X 8
10 4 X M5 Hex Nut

Fan case

11 Custom designed 3DPrinted Extruder Frame [PLA]
12 Bowden Tube Ø 4mm
13 4 X M3 X 10
14 4 X M5 Hex Nut
15 Custom designed 3DPrinted Coupler [PLA]
16 2 X 5M X 25
17 2 X M3 X 30
18 1" Aluminum Spacer, D: Ø 2"

LEFT: ABB 1600 ROBOT + TRANSLATOR 004: 3DEXTRUDER. OBLIQUE DRAWING WITH EXPLODED PARTS. RIGHT: ABB 1600 ROBOT + TRANSLATOR 004: 3DEXTRUDER. OBLIQUE DRAWING + WORM-EYE VIEW DRAWING WITH EXPLODED PARTS.

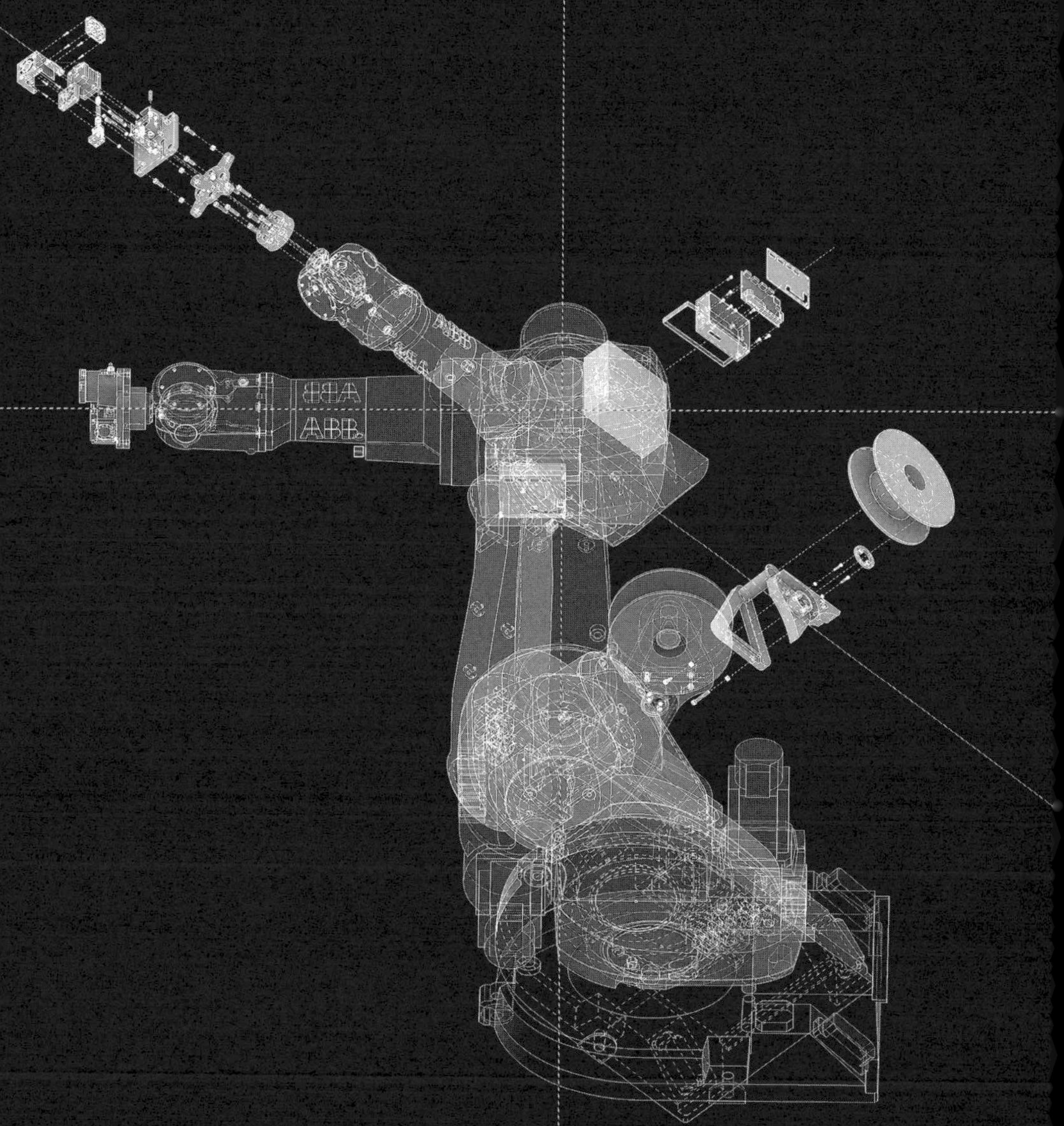

Main board
19 Velcro Strap
20 Custom designed 3DPrinted Case (PLA)
21 Main Board
22 3 X M3 X 16
23 4 X M3 X 16
24 Clear Acrylic Cao

Filament spool
25 Custom designed 3DPrinted Cap (PLA)
26 2 X M3 X 16
27 2 X M5 Hex Nut
28 Custom designed 3DPrinted Support [PLA]
29 Custom designed 3DPrinted Spool Holder [PLA]
30 2 X M5 Hex Nut
31 M3 X 30

TRANSLATOR 005_INVERTED BLOCK GRABBER

Model: Custom made + Bosch GGS 28 CE / 30.000 RPM

Capacities: This tool is used to test a different kind of edition with the mototool. In this case, a support for the MDF blocks was designed to be attached to the working table of the workshop. In a 45-degree rotation supported by the corner, the translator allowed the robot to access different faces and angles of edition. To achieve this, two PLA objects were 3D-printed. The first part is to grab the corner of the MDF blocks. The second one is to hold the block to the table in a firm, secure way. These two parts were designed to interlock with each other.

Actions: Edit, Burr, Crave, Carve, Mill, Grind, Transform, Modify, Erode, Devastate, Corrode, Mold, Extract, Texturize, Subtract, Engrave, Buff, Drill, Ruin, Gnaw

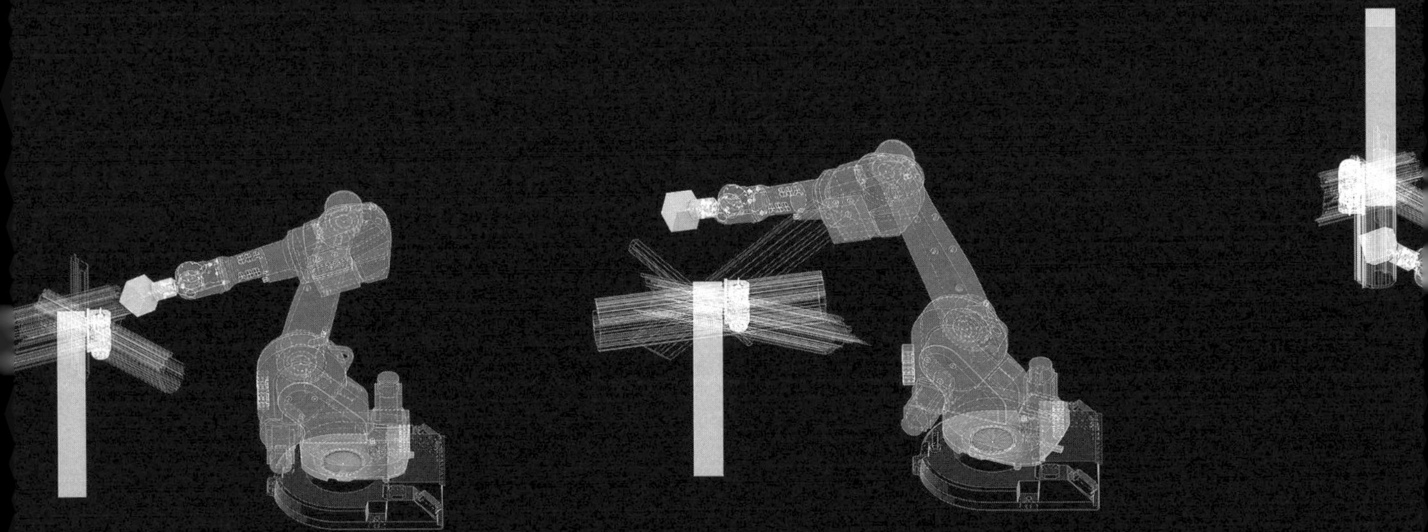

THE TRANSLATOR 005 [INVERTED CUBE GRABBER]. SIMULATION OF TOOLPATHS ADAPTED TO TRANSLATION WORKFLOWS.

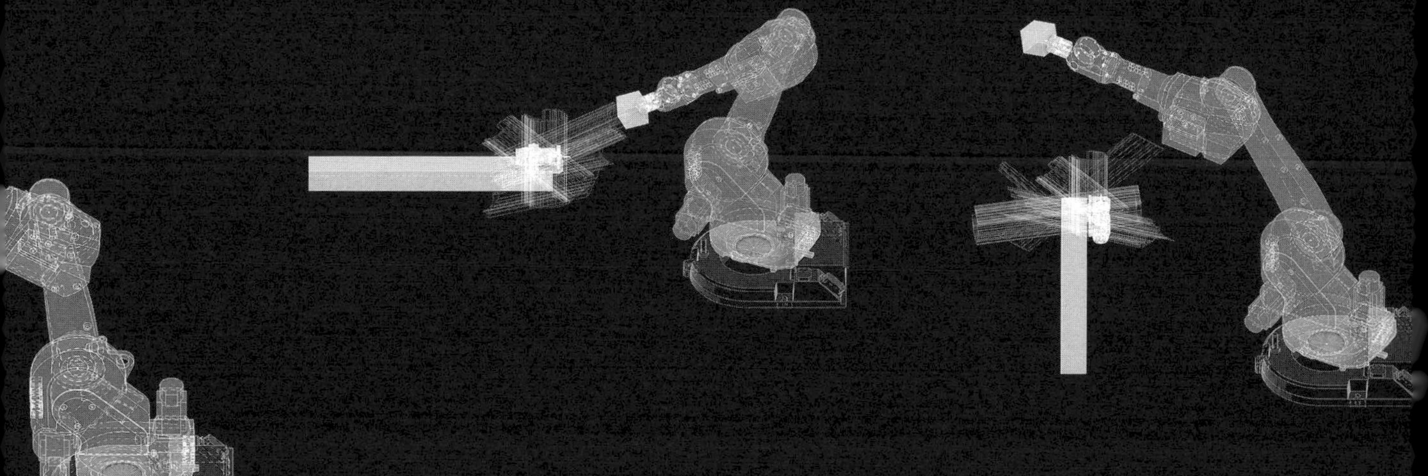

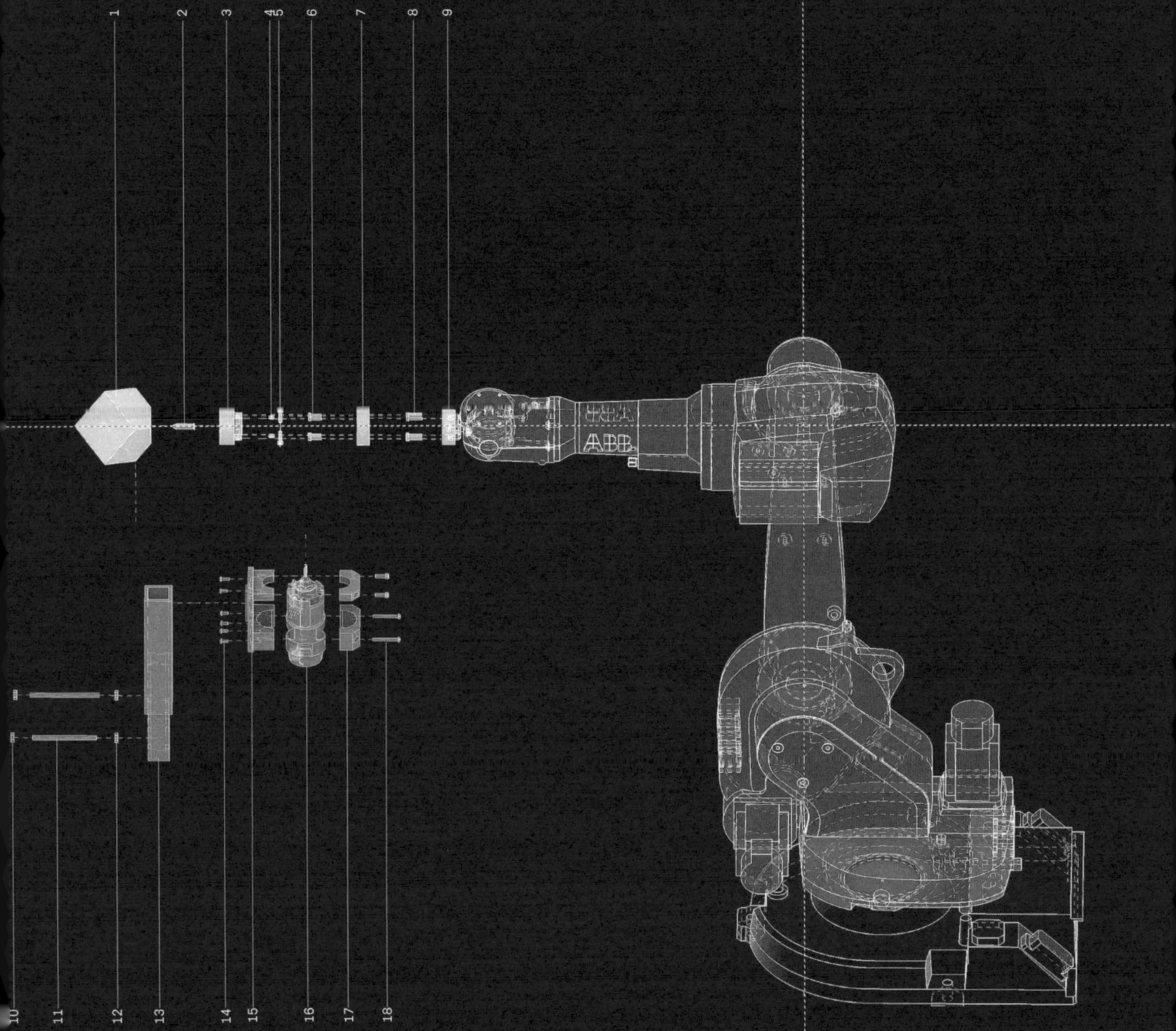

Assembly pieces

Part A. Robot attachment:

1 MDF Block
2 2 X M4 X 25 Self-Tapping
3 Custom designed 3DPrinted Cube Support [PLA]
4 2 X M6 X 25
5 2 X M6 Hex Nut
6 4 X M5 X 25
7 Custom designed 3DPrinted Coupler (PLA)
8 4 X M5 X 25, D: Ø 1"
9 Aluminum Spacer, D: Ø 2"

LEFT: ABB 1600 ROBOT + TRANSLATOR 005: INVERTED CUBE GRABBER. OBLIQUE DRAWING WITH EXPLODED PARTS. RIGHT: ABB 1600 ROBOT + TRANSLATOR 005: INVERTED CUBE GRABBER. OBLIQUE DRAWING + WORM-EYE VIEW DRAWING WITH EXPLODED PARTS.

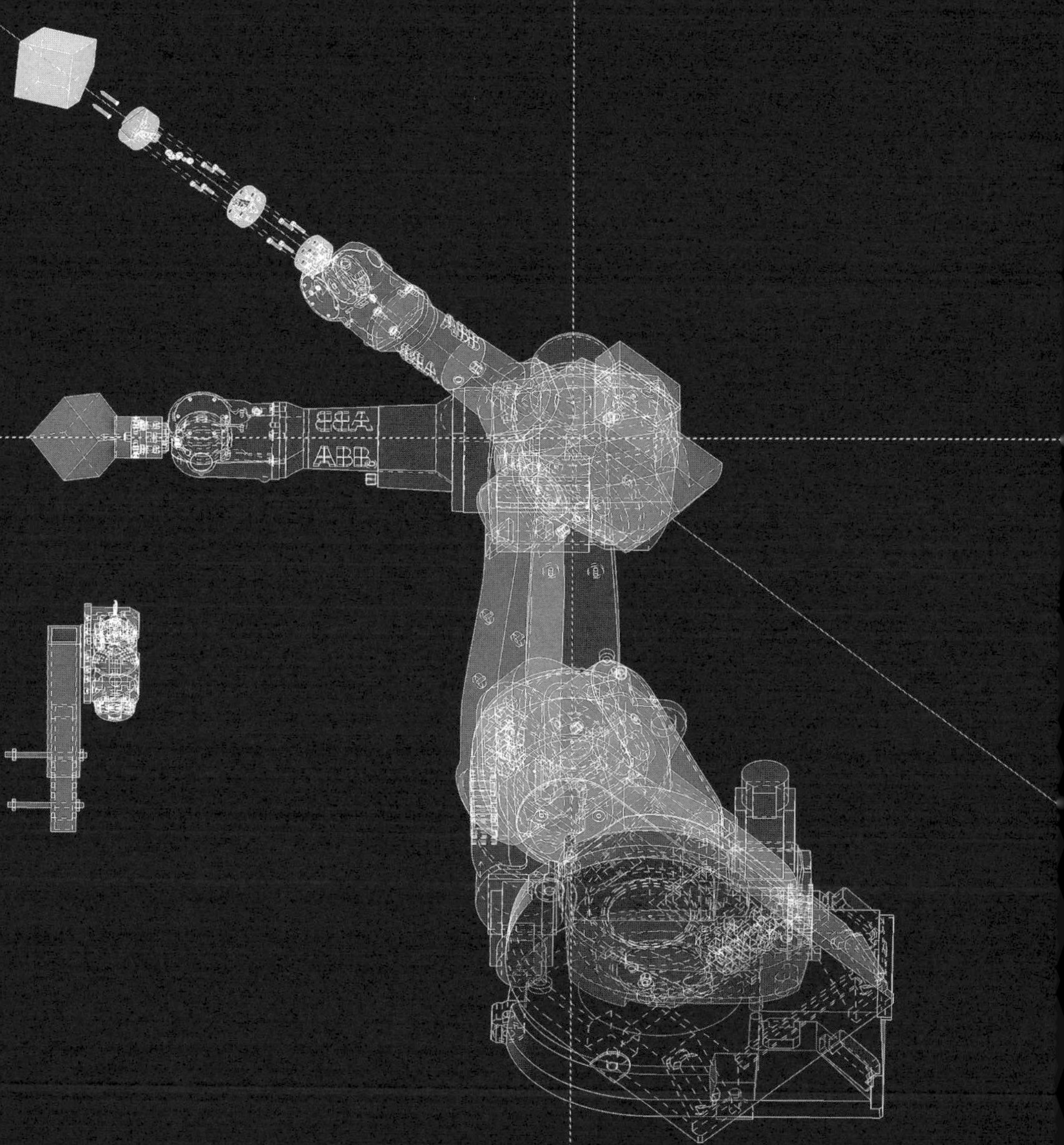

Part B. Working table adapter
10 2 X 5M Hex Nut
11 2 X M3 X 30
12 2 X 5M Hex Nut
13 Aluminium Extrusion
14 4 X M5 X 20
15 ½" Aluminum Brackets, 2 X M6 X 25 Screws [2]
16 Bosch GGS 28 CE / 30.000 RPM Mototool
17 ½" Aluminum Plate, 4 X M5 X 20 Screws [6]
18 2 X M3 X 25

COMPOSED DRAWING. BASED ON TRANSLATIONS WORKFLOW COMBINATIONS. CAMERA [OBLIQUE VIEW + EXPLODED AXONOMETRIC VIEW] + HEATGUN [EXPLODED WORM-EYE VIEW + AXONOMETRIC VIEW]

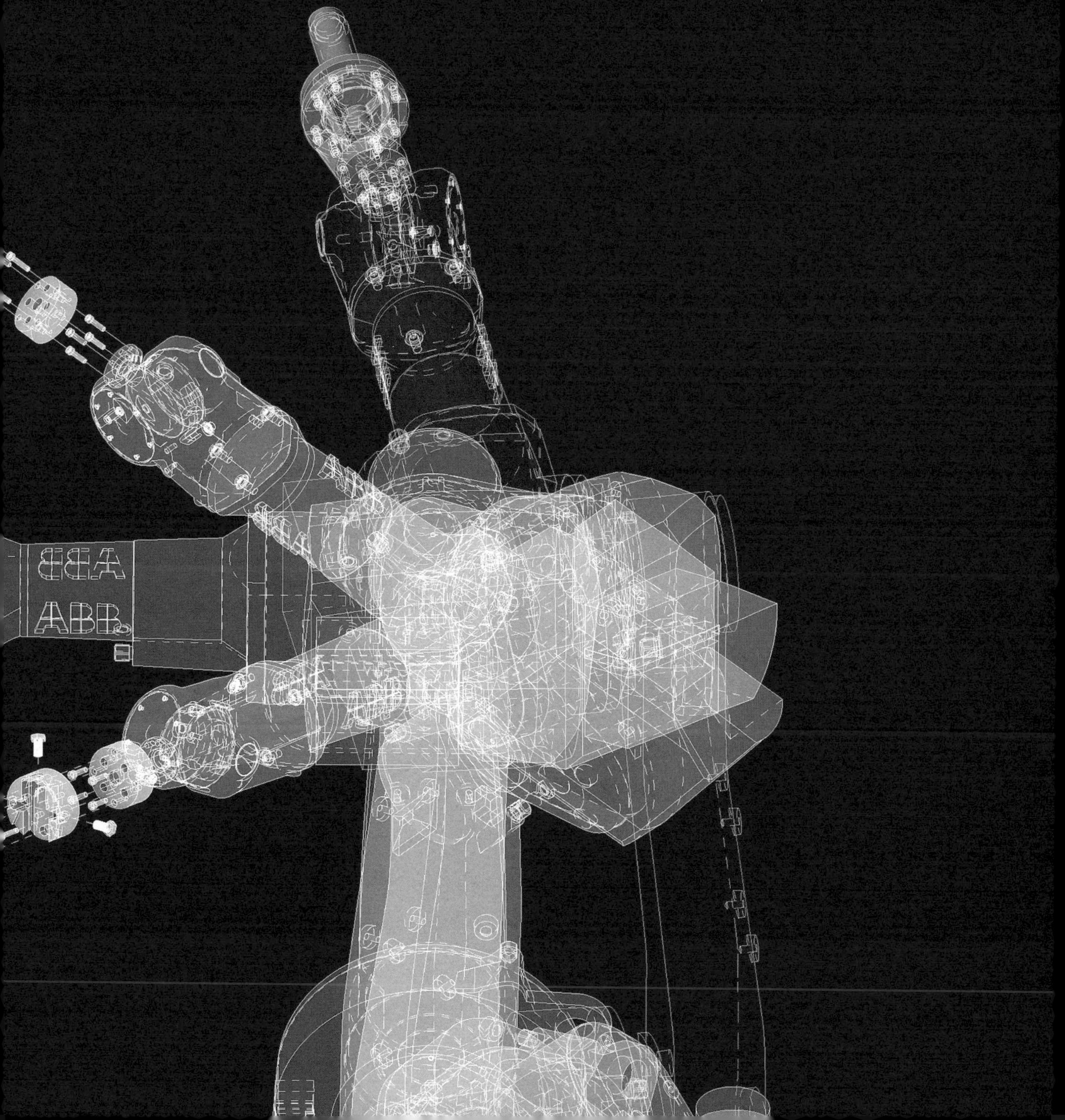
ABB

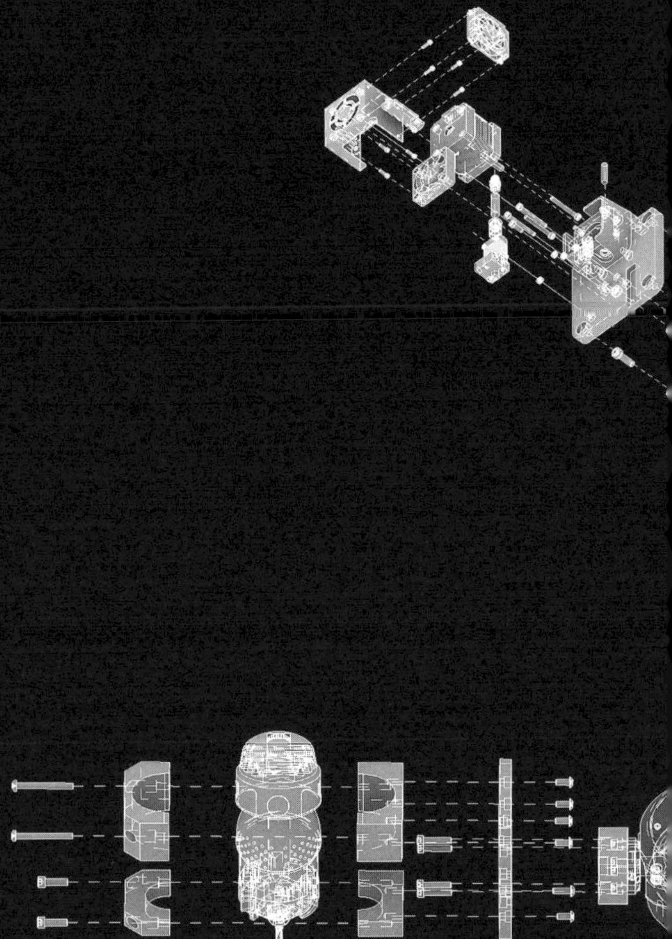

COMPOSED DRAWING. BASED ON TRANSLATIONS WORKFLOW COMBINATIONS. 3DEXTRUDER [EXPLODED OBLIQUE VIEW + EXPLODED WORM-EYE VIEW] + INVERTED CUBE GRABBER [AXONOMETRIC VIEW]

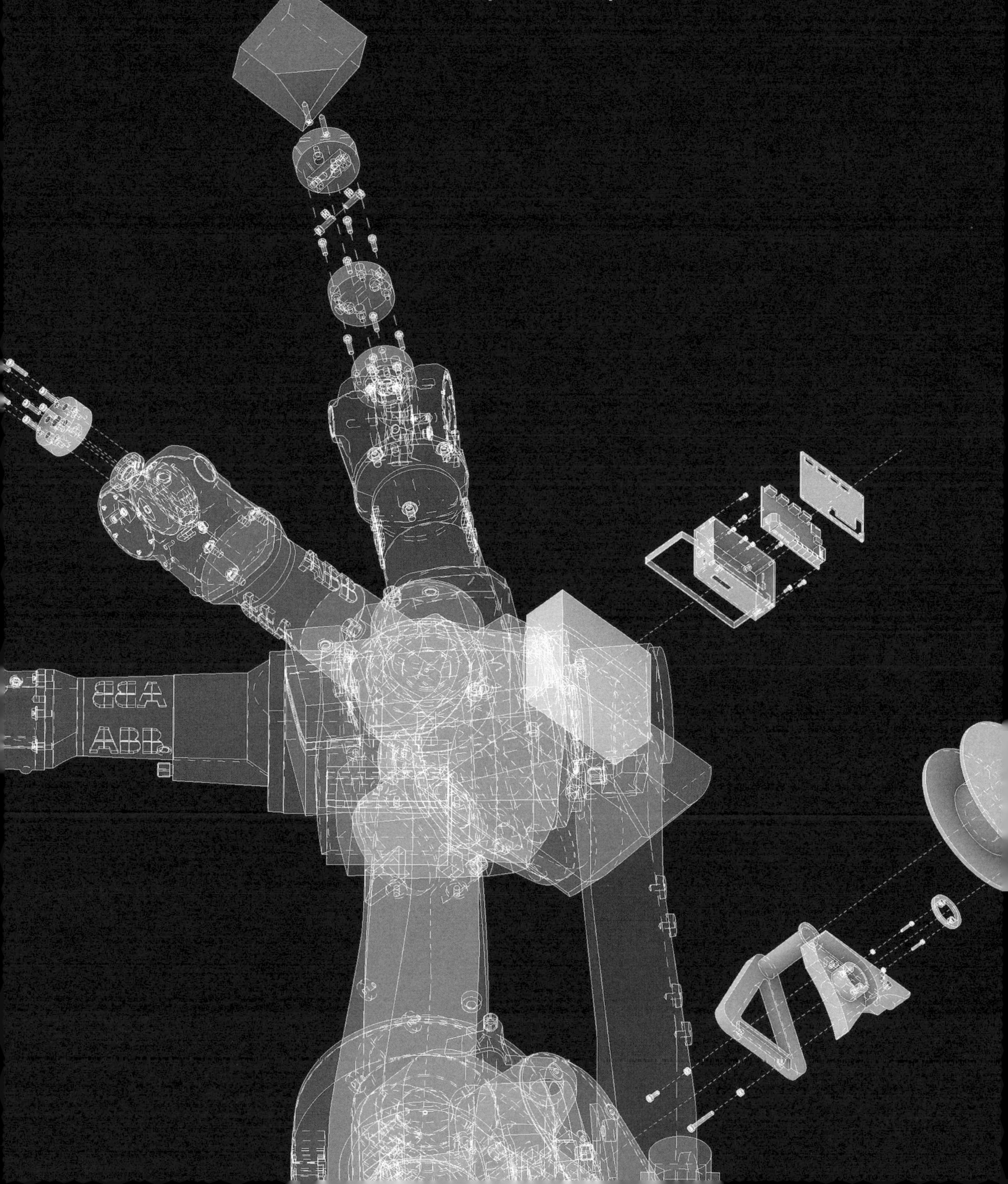
ABB
ABB

ROBOTIC CIRCUMSTANCES: DESERT Z EXHIBITION. MATERIAL CLOSE-UP DURING EXHIBITION. 2023

OFF: WHY BLACK?

DANIELA ATENCIO
CLAUDIO ROSSI

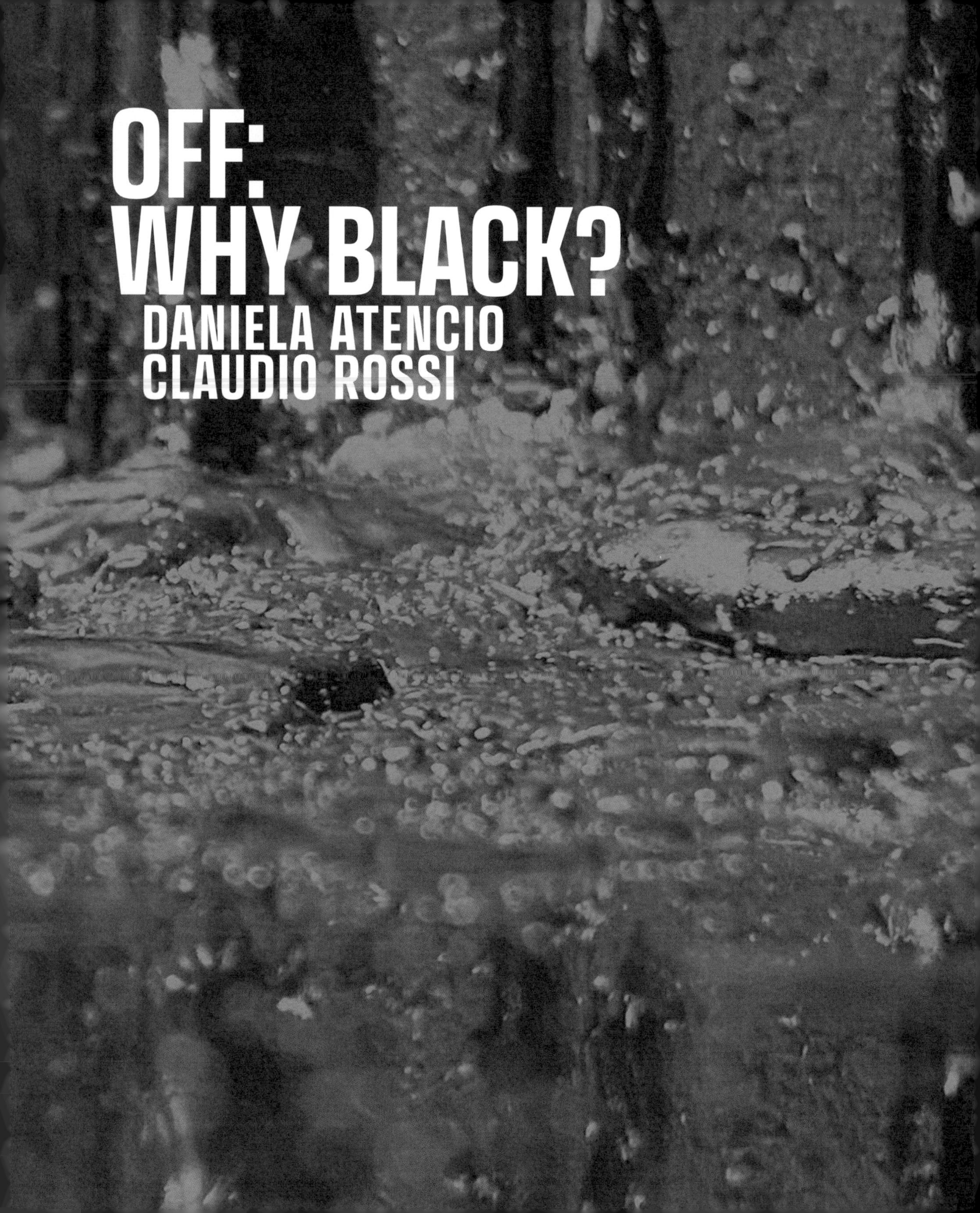

This chapter will address the construction of some parallel questions about representation that should be extended to the different processes described in the book "Robotic Translations: Design Processes." These approaches fall into two categories: The first provides a brief baseline on representation, which serves as an entry point to an ethical/aesthetic exploration accumulated in the design processes presented in the book, starting from the use of black and white *[OFF]*. This approach aligns with the narrative power of images and the descriptive nature of drawings, serving as a unique way of constructing languages and potential reinterpretations for the architectural project. The second, called procedures from representation *[ON]*, tests different techniques that promote spatial thinking in the problems studied through the use of technology, especially the robotic arm, through its choreographic performativity. In the latter, the role of technology in the materialization of physical, digital, or composite devices is amplified as a complementary transit between forms of approach and hybrid techniques accumulated in the last three years.

Base-line: *[OFF]*

In recent years, analyzing what has been raised in various academic fields that focus on teaching design exercises, representation is generally seen as a practice that is the end result, residual, and disconnected from the specific purposes of the different design studios or project workshops. In other words, representation is understood as the end state of the design process.

The practice of 'cleaning up' the extensive design process leads to two misinterpretations of the very definition of representation. On the one hand, understanding that the process is 'prior' to a 'final or definitive' state is an erroneously linear practice. The design process becomes, from this perspective, a continuous exercise, but with a foreseen end. Once completed, one speaks of a state of neatness or 'final representation and presentation'. In this sense, although they are semantically close, representation and project presentation entail different ontologies.

Both are part of the communicative skills of most of the students in architecture schools, at least in our region. They are even connected in pedagogical terms, with oral presentation or writing skills. Representation is a distinctive and differentiating element of the specific contents of the design process. They do not entail a communicative exercise per se but a reflection on the very process of what is projected, according to Alberto Pérez-Gómez.[1]

In these terms, representation is more similar to writing, not because of its communicative capacity but because of the sum of elements, parts, and systems (syntax) that are built from language. That is to say, writing as a practice of language expression, full of meanings and keys that allow us to give deep meaning to what we express.

On the other hand, it should be noted that the process and the result are different issues. Representing is in itself a continuous exercise of materialization of ideas, supported by techniques and tools that should be part of the conceptual statements of each academic exercise in architecture. Each space within the training of the architect or designer must define its methodological route to establish approaches to the project from the vehicle of representation. This promotes debates and the restructuring of old practices in the teaching of architecture and design to establish contemporary positions.

For authors such as Alberto Pérez-Gómez in The Historical Context of Contemporary Architectural Representation,[2] representation, expressed in terms of its apparatuses and mechanisms, currently possesses a charge of modernity. Plans, sections, and facades are demanded as essential elements for a disciplinary reading. The author explains that, from this perspective, we confuse representation with the process of drawing up documents for the construction of buildings.

1 Pérez-Gómez, A. (2012) "*The Historical Context of Contemporary Architectural Representation*", cit. in. Ayres, P. (Ed.). "*Persistent Modelling: Extending the Role of Architectural Representation*" (1st ed.). London: Routledge.

2 *ibidem*

Precise and disciplinary documents move away from the creative capacity that representation implies. The author also extends this position by highlighting the close relationship between representation as a vehicle mediated by technique and technology.

The tools of representation are not neutral, the author asserts, they entail utilities and have an impact on the way we see and manifest what we do. Mario Carpo would add that tools constitute what we do and how we do it.[3]

The floor plan drawings are, perhaps, the most conventional documents of the design exercise. This conventionality is based on their constitution, using elements such as lines, hachures, texts, and symbols to 'show' the architecture seen from above, with no room for interpretation. Its precision is such that the reading of these documents becomes a strictly disciplinary exercise, accessible only to those with specific knowledge in the area.

The spatial exercises of Andrea Pozzo presented in his treatise on art and architecture[4] propose a change in the gaze on representation since the Renaissance period. Pozzo dislocates the gaze upward, inviting us to think about the ceilings or the observation of domes and vaults, and constructs representational apparatuses that promote contained spatialities from habitable space. Later in this chapter, we will return to this gesture of "looking up" and its relation to robotic choreography.

Plan drawings also lead to a static situation of things or objects in space and their functions, which makes them more conventional documents. Furniture, levels, depths, and heights remain part of a hierarchical exercise (from above) and a form of presentation specific to the discipline.

Perhaps the section contains different capabilities to describe and represent spatial explorations in architecture. Authors such as Lewis, Tsurumaki & Lewis have intensified in *Manual of a Section*[5] the search for the section as a tool for research and architectural characterizations. The authors seek to establish specific qualities that allow them to organize projectual strategies through re-drawing iconic buildings. Others, such as Design Earth, have sought definitions from other fields of knowledge that allow them to expand explorations in architectural representation, such as the aquarium cut in Geostories.[6] Finally, the exercise of black and white lines in the Graphic Anatomy books[7] by Atelier Bow Wow, where sections and floor plans acquire spatial depths and are accompanied by technical definitions while sketching the different activities that take place in their projects.

This text proposes, then, to 'outline' a series of procedures to connect the approaches presented as a starting point. Each of these procedures has, in its own field of application, techniques, tools and ideas behind them, which allowed us to build perspectives that are open to the discussion on representation. They have an exploratory tone and divergent but complementary theoretical frameworks. In the end, each of these approaches proposes a recipe of its own to provide those who read this text with arguments and techniques to put aside or reconsider some issues of our discipline *(OFF)* and explore others, sometimes new, sometimes forgotten *(ON)*.

3 Carpo, M. (2014), *Breaking the Curve: Big Data and Design*, Artforum International, 526, p 169-173.

4 Pozzo, A. (1693) "*Perspectiva Pictorum et Architectorum, Vol. 2*". Roma: Typis Joannis Jacobi Komarek.

5 Lewis, P., Tsurumaki, M., Lewis, M. (2016) "*Manual of a Section*". New York: Princeton Architectural Press.

6 Goshn, R., Jazairy, EH. (2020) "*Geostories: Another Architecture for the Environment*". Barcelona: ACTAR.

7 Tsukamoto, Y., Kaijima, M. (2007) "*Graphic Anatomy Atelier Bow Wow*". Tokyo: TOTO Publisher.

Switch *OFF/ON*: or Why Black?

In this process, color variables are deactivated *[OFF]* to construct artifacts represented in black and white. The premise is the construction and superimposition of layers that inform the drawings without resorting to color as a complementary element of the visual exercise.

This involves two parallel processes: the graphic description and the composite hyper-artifacts.[8] The first, which we will develop below, has to do with the conceptual approach to description as an exercise in representation. The second, as the combination of different layers of information that come from various digital techniques and physical information that are compressed and composed into a single drawing.

The word 'describe' comes from the Latin *"describere"*, which means *"to write about how something is perceived. To explain in detail"*[9]. In this sense, graphic description is understood as the exercise of reading and representing the various details that explain what is observed or projected, that is, reality or imagination.

Ludovico Quaroni explains in his text *Proyectar un Edificio: Ocho Lecciones de Arquitectura*[10] that there is a kind of uniformity without quality in modern representation. The author, being a great advocate of color (including black), raises a key concern when he writes: *"The projects that come out of the faculties of architecture have form, but no color, just as they are not clear with regard to their component materials and therefore to their reality even technological."*

It is perhaps from this statement that we can raise certain academic concerns and some methodological proposals. Black and white drawing, conceived from these precepts, approaches a more descriptive reflection of the represented object and mobilizes the complexity towards the artifact and the means necessary to express this reality. In other words, we must look for ways to bring representation closer to the more materialized state of architectural ideas.

Instead of selecting colors to explain and, often, applying materials to an illustrative drawing, we are interested in the drawing that describes the constructive, material and technological complexity with which the architecture proposed in each of the academic exercises presented in this book is built. Lines to express construction joints, material expansions, technical arrangements, structural joints, and carpentry are just a few examples. All elements, parts and their compositions must be drawn in the exercise, developing the complex anatomy of what is projected.

We recall the illustrations of descriptive anatomy treatizes in order to approach the description. Nineteenth century anatomists sought to describe the complexity of the anatomical systems of living beings and built complex and composite representational apparatuses based on this reality. How to we show the hidden systems of the human body? How do we represent the failure of any of these systems? And even more complex, how can we demonstrate the correspondence between healthy and diseased systems in a single representational apparatus? Real, without distortions or interpretations.

Black is, in its essentiality, the 'off' chromatic element, devoid of qualities by itself.[11] To give it properties, it is necessary to attribute to it thickness, brightness, depth, contrast, or a frame of differentiation. The black line is nothing until it intersects with one and another and another, with different thicknesses and defined continuities. Full lines, over dotted, dashed, of different separations, construct a language similar to typography. These accumulations of information are reminis-

8 See *Hyper-Artifacts*

9 De Vaan, M. (2008) "*Etymological Dictionary of Latin and the other Italic Languages*", Leiden: Boston.

10 Quaroni, L. (1977) "*Proyectar un Edificio: Ocho Lecciones de Arquitectura*". Madrid: Xarait Ediciones.

11 For the Oxford Dictionary. *Black* is: "*Of the darkest colour possible, that of soot, coal, the sky on a moonless night in open country, or a small hole in a hollow object; designating this colour; (also) so near this as to have no recognizable colour, very dark.*" Oxford English Dictionary, s.v. "'black-on-white' in black (adj. & n.), sense P.11," February 2023, https://doi.org/10.1093/OED/1988892491.

cent of the work of Gertrude Goldsmith (Gego) in her *Dibujos sin Papel* (1976). The artist suspended metal wires of different thicknesses and breaks (lines) to construct dynamic, complex, and emotionally charged spaces through the use of light and black shadows projected in space.

On this same basis, the American painter Ad Reinhardt, known for his *Black Paintings* (1953-1963), who in his work renounced color, said: *"The illusionism of the black paintings is involved, ironically, in that they are non-chromatic, with a color experience. Since black represents the absence."*[12] For the artist and poet, the colors in the chromatic palette are *"barbaric, unstable and primitive."*[13] But black is something that goes *"beyond color, form [and] line."*[14] Black has a negative presence that dematerializes as a stable *"non-being"*[15] from its reflective capacity. Accordingly, the brightness of black possesses an unstable quality and an enveloping and atmospheric capacity. This leads us to two possible reflective exercises: on the one hand, to understand that black is linked to concepts such as shadow, gloom, darkness, silence, and opacity (even death). On the other hand, that black possesses a reflective capacity, that is to say, a light of its own.

Let's start with the last one, reflections on black and reflections of black. According to authors like Anhek Yilmaz buildings such as the Seagram Building (1958) by Architect Ludwig Mies van der Rohe and the Guthrie Theater (2006) by Architect Jean Nouvel, can be taken as examples of the reflective atmosphere that black possesses.[16] This atmospheric effect in buildings, at least from more conceptual and less sociocultural positions, is opposed to the modern look of the 'white buildings' designed by Le Corbusier. The architect expressed *"whiteness is the eyes of truth"*.[17] The white box would become the space ready to receive and exhibit,[18] while the black building is the container of the dark and creation.

Another example is the (self-photographed) photographer Zanele Muholi who traverses this logic in her work *Hail in Dark Lioness*.[19] The book, builds a series of portraits that manifest the reflective condition of black from the aesthetics of brightness and contrast, working on the photographer's skin from the handling of light, printed in black ink on glossy or silver papers.

This condition of black and its closeness to brightness or opacity (visually and conceptually), presents a question about the dioptric (seeing through, i.e. transparency) and the catoptric (or the capacity to reflect) of what we see, and ultimately, of what we represent. Black is opaque and reflective depending on the type of observation made. In fact, it is both, as Alexander Galloway refers to in *The Black Universe*, referring to *Reza Negarestani Cyclonopedia: Complicity with Anonymous Materials*,[20] when he takes as an example the metaphor of oil: *"it is black, of course, in color but only in its deplorable morality. But it is also*

12 Reinhardt, A. (1970) "*Black Paintings, 1951-1967*". Edited by Rose, B., & Arnason, H. New York: Marlborough Gallery, Inc.

13 *Ibidem*

14 *ibidem*

15 *ibidem*

16 Yilmaz, A. (2016) "*Color of Absence and Presence: Reconsidering Black in Architecture*". COLOR Research and Application. London: Wiley

17 The authors refer, in this case, to the differentiation between the modern museum (the container) and the *black-box*. See: Mostafavi, M., & Leatherbarrow, D. (1993) "*On Weathering. The Life of Buildings in Time*". Cambridge: MIT Press.

18 *ibidem*

19 Muholi, Z., Mussai, R., Azzua, U. (2018) "*Zanele Muholi: Somnyama Ngonyama, Hail the Dark Lioness*". New York: Aperture.

20 Negarestani, R. (2008) "*Cyclonopedia: Complicity with Anonymous Materials (Anomaly)*". cit. in. Galloway, A. (2014) "*Laurelle Against the Digital*". PostHumanities 31, Minneapolis: University of Minnesota Press.

light because it is a transmutation of sunlight. Oil is a geological product of sunlight, first by photosynthesis with vegetable matter, and second, by the decomposition of vegetable matter over time."[21]

Galloway, in a discussion of Laurelle's ideas, proposes to deepen this reflection on blackness by coining the phrase: *"we must leap not from light to darkness, but from darkness to 'blackness'."*[22] This phrase refers to the search for answers to reality from the blackness of science. The true vision of reality, or that closest to the truth, is the one that can be obtained by being in a dark room, looking for small reflections or glimmers amidst the opaque and lightless totality.

Darkness, blackout, blackness, shadow, gloom have been some of the connotations attributed to black. From the narrative point of view, the representation determines symbolic traces that allow establishing connections. The position on black and white, from light and darkness recalls the observations of Galileo or Newton explained by Frank Wilczek,[23] on the contrast in the observation of the lunar sphere. The white orients through its reflection, the black, referring to the 'dark side', moves curiosity. The balance that arises from this contrast, by opposition of their qualities, manages to amaze those who have observed the sky.

In Lawrence Stern's novel *The Life and Opinions of Tristram Shandy, Gentleman*,[24] the author uses completely black pages to determine obscure narrative points in the central story. While not legible, these pages mark significant moments in the development of the novel. The way these pages can be understood is as the priming of a lot of information repeatedly placed in different positions, until the whole is covered. It is the pounding of the typewriter type on the white sheet of paper over and over again. In this sense, we can say that black is the construction from the superposition of information and not the simplified idea of an empty totality.

According to Yilmaz, this is reminiscent of Kazimir Malevich's *Black Square on a White Background* (1913), or *Shadows on Night Hunter* (1955). According to the author, Malevich added: *"The familiar recedes ever further and further into the background.The contours of the objective world fade more and more and so it goes. No more 'likeness of reality', no idealistic images-nothing but a desert. But this desert is filled with the spirit of non-objective sensation, which pervades everything."*[25] In this sentence Malevich talks about non-objectivity and this desert atmosphere close to opacity. These non-idealistic images are part of Jason Payne's analysis in *Variations on the Disco Ball, or the Ambivalent Object*.[26] They refer to a condition of mediated observation and accumulated information.

Black is not to *"please the eye,"* according to the painter Odilo Redon.[27] Black is an element that builds an atmospheric reality more meaningful than the color palette. It is in this specificity and in opposition to the breadth of the color palette, where black, by opposition and contrast, determines specific characteristics of the objects represented. For the philosopher Maurice Merleau-Ponty in the incomplete manuscript The Intertwining-The Chiasm, *"our perception of color is constantly and simultaneously being modified*

21 *ibidem*

22 Galloway, A. (2014) "*Laurelle Agaisnt the Digital*", PostHumanities 31, Minneapolis: University of Minnesota Press.

23 Wilczek, F. (2016) "*A Beautiful Question: Finding Nature's Deep Design*". London: Penguin Books.

24 Stern L. (1790) "*The Works of Laurence Stern: The Life and Opinions of Tristram Shandy*", Gentleman. London: Proprietors.

25 Yilmaz, A. *Op. Cit.*

26 Payne, J. (2013) "*Variations on the Disco Ball, or, The Ambivalent Object*", Project 2 (Summer), p.20-27.

27 Yilmaz, A. *Op. Cit.*

as a supplementary experience of light, texture and form".[28] From this observation, it is in the continent of the materialized description of objects and their elements, where the specific content of black and white representation resides.

Black is, from this procedure, a pictorial tool like those of Reinhardt, Malevich and Redon, a literary machine like that of Stern, sculptural and phenomenological like that of Gego and undoubtedly, from Merleau-Ponty's position, a way of transferring from representation, the very materiality of reality. It is in the complementarity between black and white and their possible intermediate grays, that there are possible mediations in art such as chiaroscuro (e.g.: Caravaggio), but also the debates that have motivated philosophy or contemporary cinema. Gilles Deleuze explains this in *Cinéma 1. L'image-mouvement*,[29] when he describes the importance of black or white screens in the construction of cinematographic narratives as the beginning or end of sequences and their possible role within the context of the images that follow.

Non-Blacklist Procedures: *[ON]*

This procedure is presented as a workflow towards composite and complex results within the precepts of the process *[OFF]*. This workflow, or the non-black list, is developed in the following six steps: The Detailed Model, Linework, Constrasted Shadows, Materialism, Background and Viewpoint.

<u>Detailed Model</u>

This step is the foundation of the whole procedure. It is the work of modeling the project object in its entirety. The metaphor of the project as an anatomical body allows to change the perspective on the idea of 3D modeling. First of all, from the process itself, thinking from the parts that make up the totality, its spatial thicknesses, and its different architectural characterizations, leads to a precise and rigorous work process. This process starts from the two-dimensional elements to gradually build the three-dimensional totality.

From this anatomical understanding of the model, the structural systems (and its literal similarity to the bone system), the organs, the circulatory or nervous system, among others, form a single perfect three-dimensional body, where everything connects and fits together precisely.

Secondly, it is not enough to model parts and elements of the 'body' based solely on their anatomical properties. For this, it is necessary to incorporate anatomical characterizations into the first step of this procedure, such as their functioning, movements, hidden systems, and emergent properties. The clearest example of this explanation is the modeling of circulatory systems. From this metaphor, we could say that the three-dimensional model of this system would consist of the shape of the veins and arteries, together with other vascular, microvascular systems and the heart. However, this would not be sufficient, since only the specific occupancy or spatial conformation of it would be understood. Therefore, in the second step of this procedure, the movement of the blood, or the blood itself, must be considered to expand the capabilities of the model.

However, there is a third aspect to consider in this procedure: the modeling of the anomalous. It is a matter of identifying and representing those manifestations that are far from the ordinary, considering both the metaphor and the anatomy of the ordinary. To achieve this, it is essential to "make the invisible visible", that is, to express from the model itself and in the subsequent stages of the procedure, that which is hidden, that which is different, particular, unique, and specific. This can be manifested both in the spatiality of the systems and in the specific expressions of the anatomy.

28 Merleau-Pounty, M. (1968) "*The Visible and the Invisible*". Evanston: Northwestern University Press.

29 Deleuze, G. (1983) "*Cinéma 1. L'image-mouvement*". Paris: Les Éditions de Minuit.

Linework

In this step, each descriptive exercise modeled from each part, element or system of the previous step is enhanced by working with lines. Make2D, lines and transparencies reinforce the visible and invisible of the whole system. Layers are superimposed to determine the accumulations between the hidden and the literal. Dotted lines, or lines with different dots, change the description of the model. Accumulating, overlaying, and hiding are essential actions. Thickening or thinning lines constitute precise actions on the model and the way it is understood.

Contrasted Shadows

This is the exercise of spatial depth of the model. It does not have a certain scale and its projection comes from the modeling and the type of drawing developed. Chunks, Axo's, Cuts or escaped plants can be complemented by shadows in high contrast. This exercise involves a reflection from light (white), shadow (black) and penumbra (grays). It is possible that the representation device comes from the precision of the modeling (orientation, heights, depths, rigorously modeled surfaces) and the desired effects (render or linework).

From this point, we can illustrate the capabilities of black and white to create transparencies, opaque bodies, penumbras, or superimpositions of brightness. We build exponentially the way in which the anatomical body of the projected reacts to light. We determine whether it is an absorbing or reflective body and whether it is legible in the dark or only under natural light.

Materialism

In this exercise, the conceptualizations of the previous chapters, which form part of the substantial basis of the research behind this book, complexify and give capacities to representation. Time, responsibilities, processes, and material agencies make up a world of explorations that allow us to give capacities to the project and challenges to its apparatus of representation. Wear and tear, ruin, non-human engagement, are some of the terms explored in this part of the exercise. To achieve this, the work of material mapping and rendering is fundamental.

Background

This is the contrasting background that supports the drawing. Black backgrounds under white lines, shadows, and rendering. In contrast, a white background can highlight the work of the other components in black, highlighting the previous steps and their capabilities within the rendering.

Viewpoint

The viewpoint is the closure of this exercise. For the purposes of this research and its conceptual (and visual) positions the *Worm-Eye View*[30] is the desirable tool. Not only as a result of the workflow, but also as an exploratory part of a different look at the developed spaces.

This work path is presented as a "recipe" of its own and not as a paradigmatic manifesto. It is a workflow that has developed its own language over the last few years, based on personal obsessions and necessary arguments, with the sole objective of stimulating well-founded and rigorous reflections in the various academic fields where it has been put to the test.

To be clear, always in *black and white*.

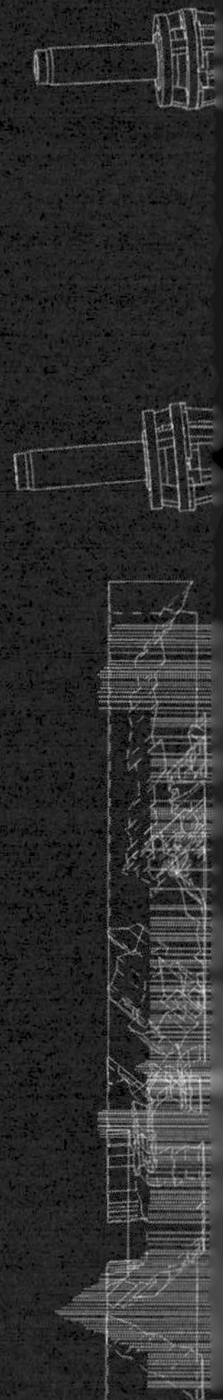

NON-BLACKLIST PROCEDURE. EXPLODED ELEMENTS IN WORM-EYE VIEW-OBLIQUE VIEW.

ABB

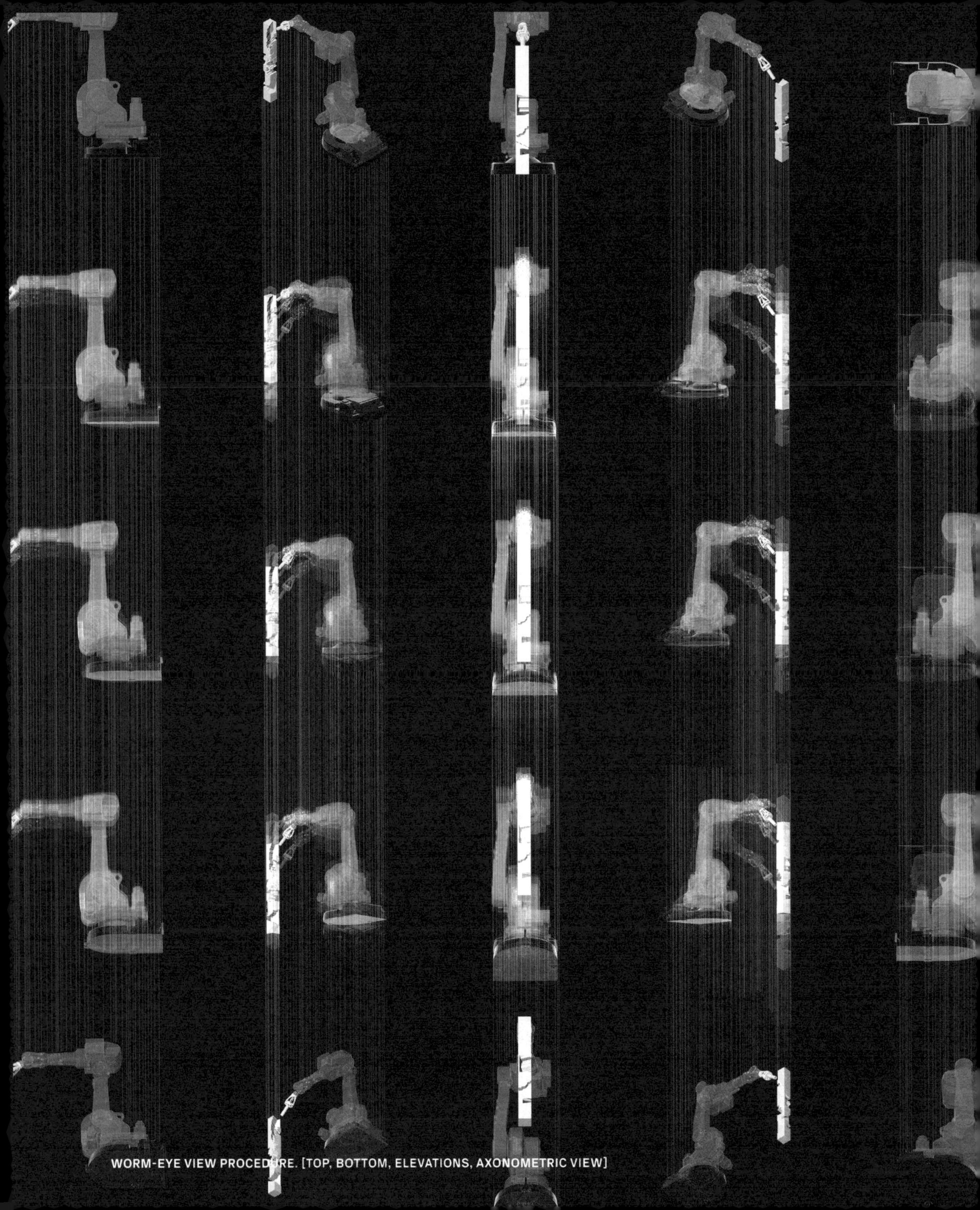

WORM-EYE VIEW PROCEDURE. [TOP, BOTTOM, ELEVATIONS, AXONOMETRIC VIEW]

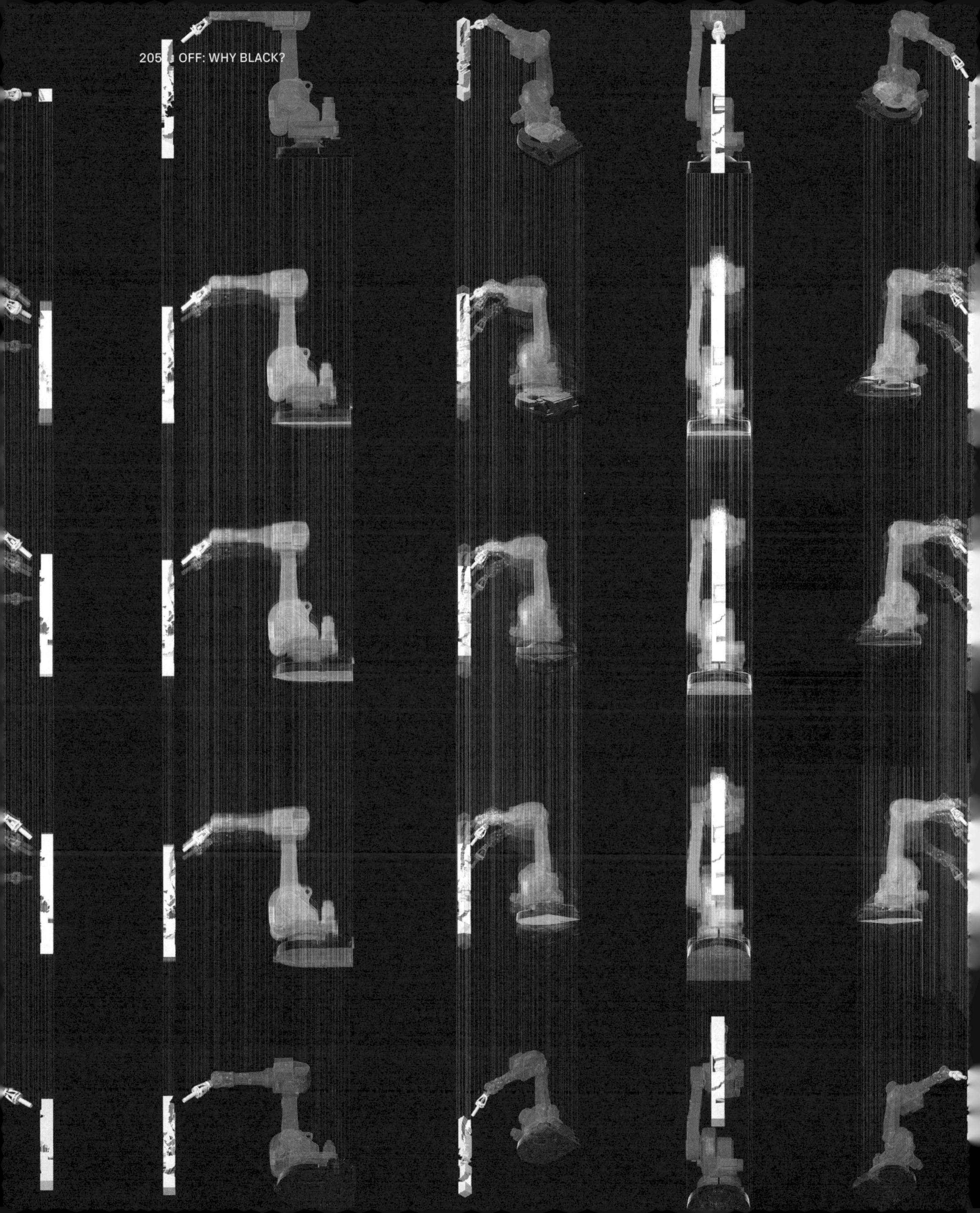

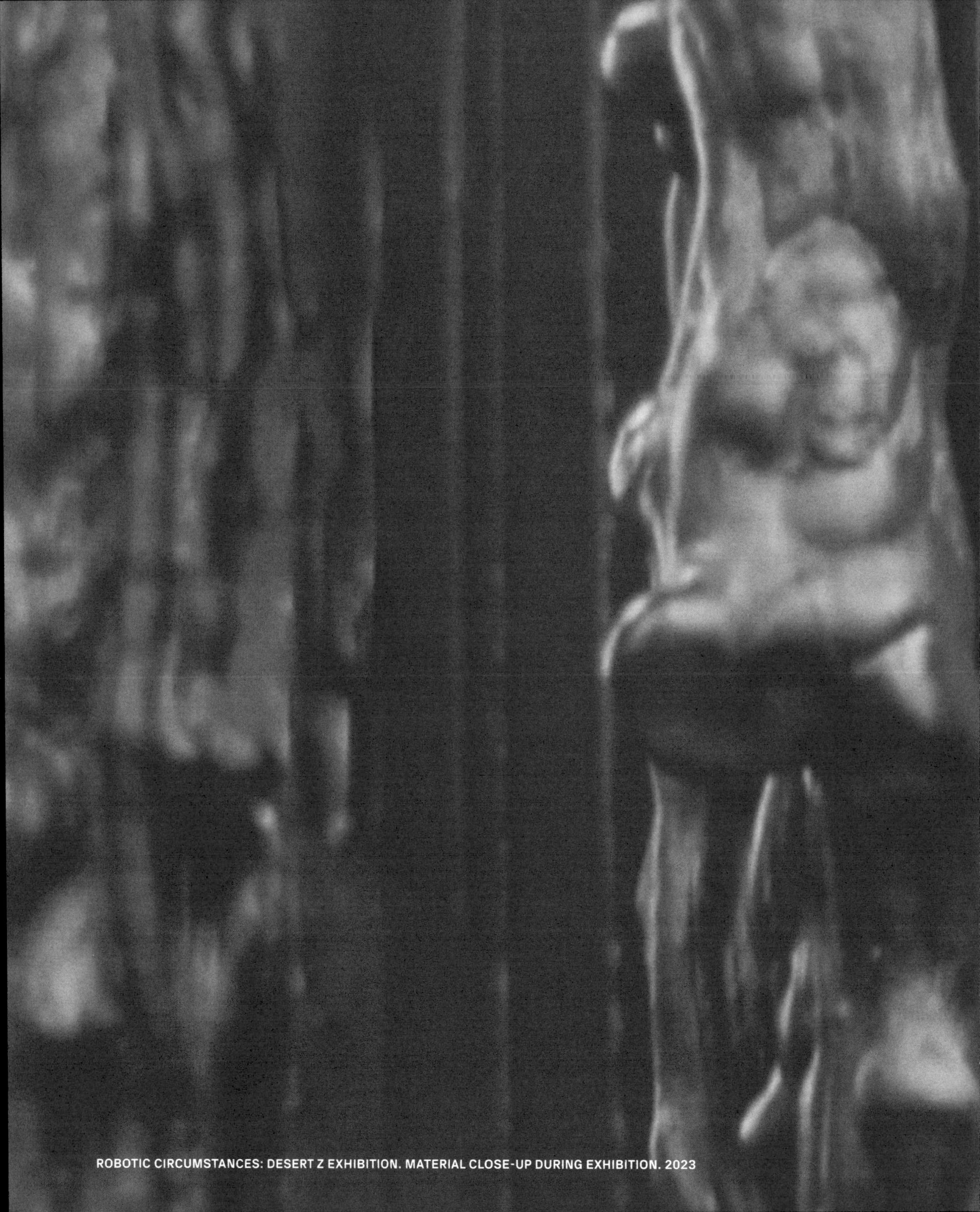

ROBOTIC CIRCUMSTANCES: DESERT Z EXHIBITION. MATERIAL CLOSE-UP DURING EXHIBITION. 2023

INTERVIEW WITH: PETER TESTA

DANIELA ATENCIO

How do you perceive the state of the art in the pedagogy-robotics relationship within architecture schools?

Robot House at SCI-Arc focuses primarily on exploring coordinated robotic motion and its potential applications in architecture. Our approach differs from productivist thinking and traditional fabrication methods. While there are various innovative approaches to using robotics in architecture schools, the prevailing trend is to employ robotic arms for direct production or assembly, similar to the linear progression seen with CNC machines in digital fabrication. Furthermore, the conventional approach often involves large payload stationary or track-mounted robots placed individually in workshops, following an industrial production model. However, this approach tends to overlook the unique characteristics of the six or seven-axis robot arm, which offers a spherical workspace and is not restricted to Cartesian models of space or predetermined operations.

In our work, we explore a new paradigm beyond the industrial model. Our work is original in shifting the focus to collaborative multi-robot workspheres and motion control interfaces. This post-production approach, which is speculative and playful, offers a contemporary pedagogical platform that bridges the digital and physical/material worlds. By merging the studio, laboratory, and worksite, we can develop new models for rethinking the relationship between design and production. This is particularly relevant as society increasingly questions the industrial model of carbon modernity.

In the Anthropocene era, it is crucial to prioritize knowledge production and pedagogy that address unpredictable factors and possible future contingencies that are on the move rather than fixed and stable. This pedagogical approach aims to equip architects as knowledge workers with new skills, allowing them to hack into and transform prevailing digital platforms and interfaces as well as older models of production and material systems in architecture while acknowledging that technology is a basic part of being in the world.

What challenges does the use of the robotic arm represent in contemporary design explorations, and with which computational processes should it be hybridized?

The user interface for robotics is becoming more open and user-friendly, thanks to tools like Rhino Grasshopper and Python. However, to fully engage and mobilize the digital/physical convergence, designers need to develop workflows that combine multiple toolsets in unexpected ways. It is now common to see designers working across different interfaces that are adapted to the material process, motion sequence, and project parameters. In our case, this hybridization began with the early use of Maya as a design, simulation, and motion control interface, coupled with computer vision and machine vision.

Digital Twin technology utilizes real data to replicate processes in both digital and physical environments, making it increasingly prevalent in various applications. Digital Doubles or Digital Twin technology is inherent in the merging of computer and machine vision with augmented robot arms. The discrepancy between the physical and digital worlds becomes apparent when working with robot systems, offering a valuable learning environment to understand the potentials and limitations of these technologies. However, these technologies still operate within a Cartesian ontology that reduces the physical material world to geometry. By incorporating sensing and feedback into a multi-robot platform, we can develop a more critical and rigorous understanding of the implications and architectural potentials of digital doubles and AI machine learning toolsets. The challenge is to meaningfully constrain these digital toolsets and workflows to advancing the project of architecture as a specific form of knowledge and praxiography.

If we assume that the robotic arm is an instrument to produce explorations from diverse creative and experiential learning processes, how would you define these pedagogical machines?

I appreciate your use of the term "pedagogical machines." To explore the theoretical framework of your question, I would like to refer to Georges Simondon's ideas on machines and technology. Simondon argues that technology is a process of invention, a continuous process rather than a fixed product. Tools are not simply used passively; they are actively reconfigured, reinvented, extended, and mutated through their use.

Personal computers (PCs) often isolate, fragment, and flatten, whereas spatial and material computing modalities affiliated with robotics promote collaborative learning by providing multiple avenues, entry points, and possibilities. Our experience has shown that working with multiple relatively small-scale robot arms enables students to safely engage with a dynamic environment for visualization and creation. We have observed that this innovative design and imaging platform enhances and accelerates collaborative learning. Speculative thinking in architecture has always been linked to new ways of seeing, and this also applies to the use of robotics.

Representation is the assembly of information. It can be understood as descriptions of parts that collide between their specific role for the whole and as autonomous and accumulated thicknesses of meaning (Latour, 2007). What new conceptual approaches are derived from representation and its relationship with technology to build these thick apparatus?

The interest in rethinking representation is related to our previous discussion. Firstly, it is important to acknowledge that the construction of apparatuses has a long history in architecture. I am particularly interested in exploring architecture as a unique form of representation and finding new ways to engage in this disciplinary discourse.

Bruno Latour's concept of "thick materialism" is relevant as it challenges an idealistic view of matter that obscures the potential of post-digital techniques and material practices. Instead of providing functionalist and utilitarian explanations for technological change, the focus on "Thick Things" examines how objects can serve unexpected purposes beyond their original design.

Ortega y Gasset explains in Meditaciones del Quijote (1914) that everything should be understood from circumstances. The Spanish philosopher refers to two definitions from this consideration. On the one hand, from the etymology ("circum-stances") as the comprehension of the surrounding stances as a spatial representation of reality. On the other hand, from its temporal meaning, the circumstantial leads to the materialization of time. These two approaches recall the spherical space (stance) of the robot and the temporal dimension of the robotic choreography or the real-time representation of it. Remembering Ortega & Gasset, do you think the robotic arm is a circumstantial machine?

Ortega y Gasset's critique of positivism is prescient and considering ontological questions may provide an alternative perspective on the industrial robot. It can be seen as a circumstantial machine, operating as an open system rather than a closed one. This perspective also applies to its interactions with the environment, both with humans and non-humans. There are still many mysteries and unknowns to be explored that are not apparent within the conventional model of digital fabrication and robotics, which tends to focus on cause-and-effect relationships.

As you mentioned, the concept of time and the circumstantial is inherent in the spherical geometry and motion space of for example the Staubli robot. This problematizes a whole series of assumptions regarding the homogeneous space of digitality. This apparatus mobilizes space and time differently. As I have previously argued the enactment of an immanentist viewpoint, one that is placed within the motion/material construct opens the possibility "to think not about motion but according to motion."

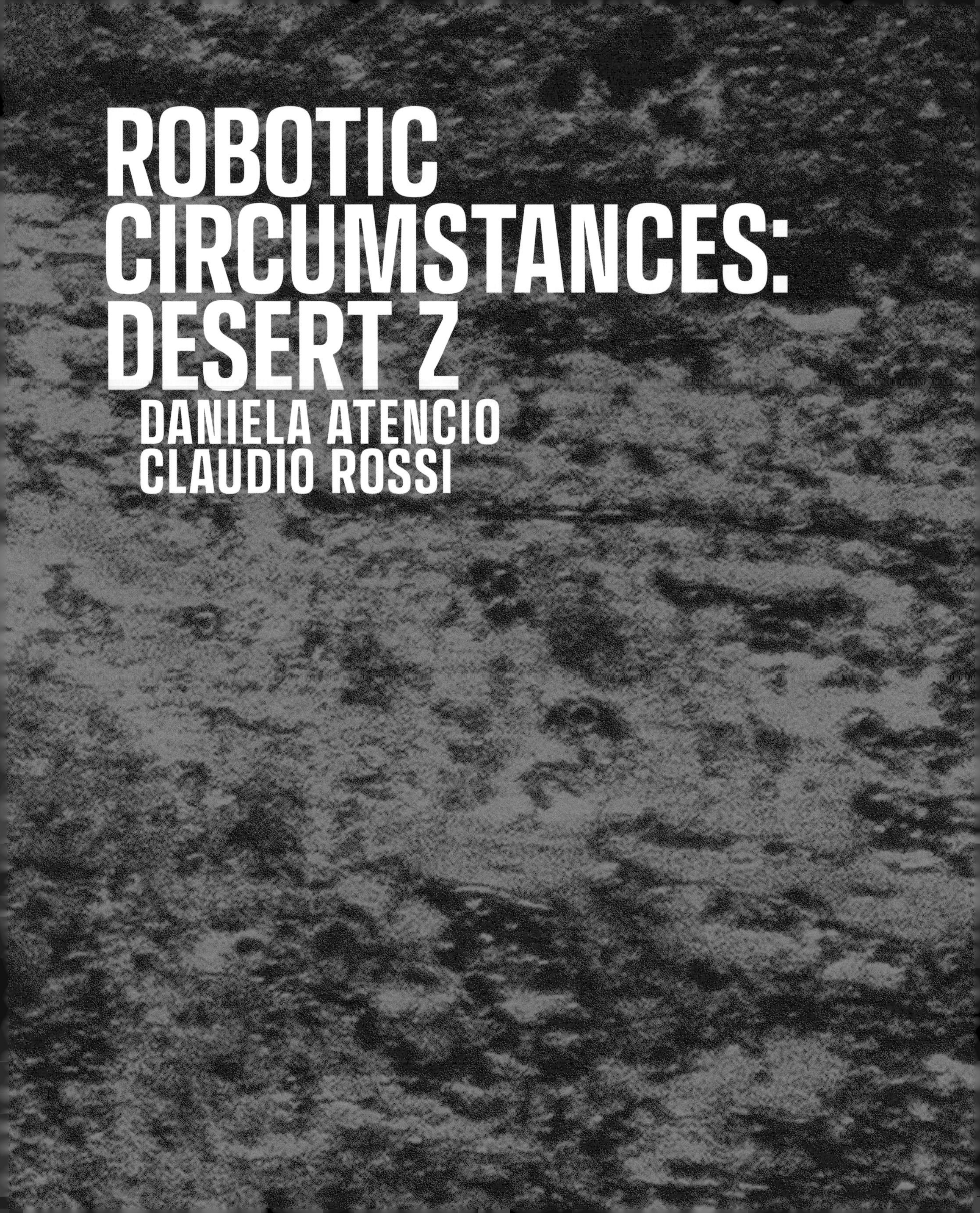

ROBOTIC CIRCUMSTANCES: DESERT Z

DANIELA ATENCIO
CLAUDIO ROSSI

DESERT_Z exhibition was a circumstantial materialization of the Guajira desert from two approaches: The first is from the translations of temporal and spatial processes expressed with the robotic arm. The time in which the pieces move and transform (chronologically) during the exhibition conforms to a set of precise and performative changes and effects.

The robot edited the material through heat using a custom-made heating blowing tool, creating a double reading: The first, the heat of the desert, its population, and its culture, but also the 'heat' as a global climatic change effect —and its affectation, that is inescapable and unstoppable. The results revealed a hidden transformative and performative landscape over time. The second reading was from the circumstantial space of the robot reminiscent of the *Pioi* for the *Wayuú's Yonna*[1] dance.

As proposed by *Ortega y Gasset* in *Meditaciones del Quixote*[2], space is circumstantial insofar as its etymology alludes to the stances that surrounds the forms of inhabiting it. In this sense, the robot, and its circumstantial space of action (360 degrees) promote a surrounding and a temporal context between material agents as interpreters (vertical witnesses[3]), and results interpreted as new material capacities and behaviors. The other approach from its etymology is that circumstances are affected-defined by time. The circumstantial space of the exhibition and the robotic toolpaths were programmed to develop a specific journey with a controlled 'dance'.

The robotic choreography was programmed to execute the edition of the 12 vertical 'witnesses' in a 3-day exhibition with 6 sessions. Each 30-minute editing session represented a time compression studied with precision from the material behavior. Each 30-minute session equals 1800 seconds. Each second was translated (metaphorically) into a day, accumulating a total of 30 hypothetical years during the exhibition. Desert_Z asked about the past, the convulsive but beautiful present, and a possible future landscape transformation by the year 2053.

The witnesses were placed around the robot as twelve pieces aligned to the chronological time marks of the clock but also oriented in a specif-

1 Carrasquero, A., Finol, J., & García, N. (2009) "*Símbolos, Espacio y Cuerpo en la Yonna Wayuu*". Revista de Ciencias Sociales, 15(4), 635-652.

2 Ortega y Gasset, J. ([1914] 1995) "*Meditaciones del Quijote*". (3.ª ed.) Madrid: Cátedra.

3 For this exhibition the vertical wooden blocks were named 'vertical witnesses', since its role was to 'witness' the material transformation through the time of the exhibition by the action of the heated air with the specific toolpath of the robotic choreography. Alaso 'vertical' since each were standing and changing the point of view of the landscape, not from above, but from the way the robot 'sees.' Also refer to: *Latin-America in Worm-Eye View.*

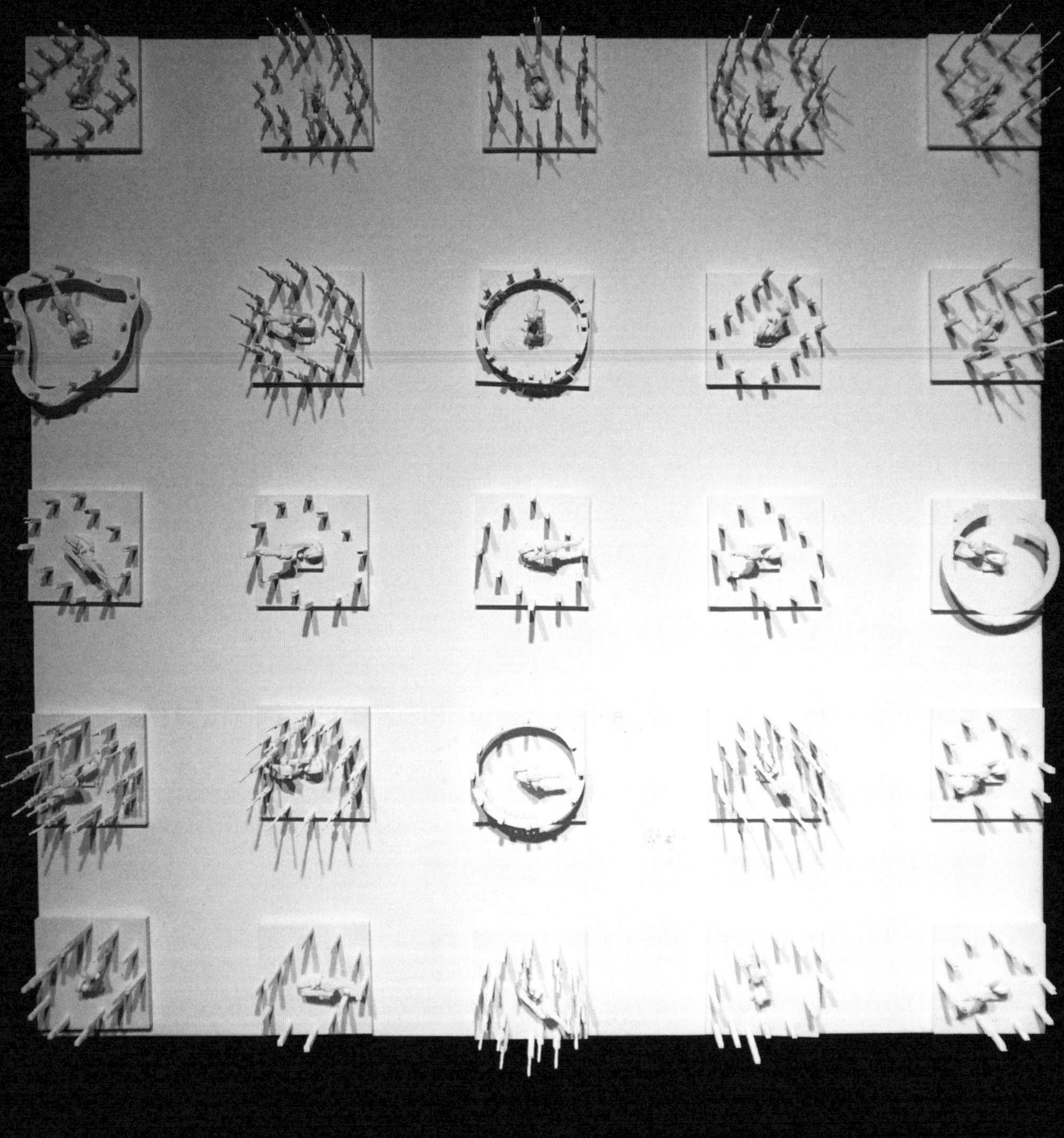

ROBOTIC CIRCUMSTANCES: DESERT Z EXHIBITION. DESIGN ITERATIONS PHYSICAL MODELS [PROCESS] PRESENTED DURING EXHIBITION. TOP VIEW. FABRICATION: 3DPRINTED + PAINTED. 2023

ic cardinal orientation. Each 'witness', and from the application of the workflow proposed by this research-creation, translated the observation of the Guajira desert into geometries that were edited in the block mass. In other words, the landscape was excavated in each block from the collision of geometries captured by the process of analogue translation and projected as 360 degrees (evolving motion).

After having the pieces of wood edited, they were filled again (to reach the original shape of the blocks), with black paraffin. After multiple material tests, some more sustainable than others, this material was selected for two reasons: first, for its ability to transform, i.e., for its behavior and expressiveness. Second, because it is a material derived from the petrochemical process of oil (petroleum) and this region is part of the oil plain of the Gulf of Venezuela. The guajira is and has been colonized by ancestral extractionist practices; therefore, this material has a degree of material criticism.

From these variables, the toolpaths were timed and designed to go through a process of material editing with the action of heat, understanding the density of the paraffin — different mixtures were tested several times. Stop, go on, move to another witness, go back, and melt more were choreographic actions thought for the exhibition in a precise way.

The result was an unexpected act of diverse narratives. Wonderful and unsuspected transformations in the 'witnesses' allowed us to see the manifestation of the conceptual starting point of this exhibition through the different material behaviors, spatial expressions, and interactions with the visitors (from the smell of melted oil to the temperature of the room), which made this experiment the perfect closure for this research-creation journey.

The landscapes hidden in the mass of the 'witnesses' were appearing, and the materiality that hid them was disappearing, as a kind of archeology on the most structural base of a landscape: its own geography.

LA GUAJIRA DESERT. OCTOBER 2022. LANDSCAPE TRANSFORMATION / EVOLUTION. INFORMATION FOR DESIGN AND ROBOTIC TOOLPATHS.

LA GUAJIRA DESERT. APRIL 2023. LANDSCAPE TRANSFORMATION / EVOLUTION. INFORMATION FOR DESIGN AND ROBOTIC TOOLPATHS.

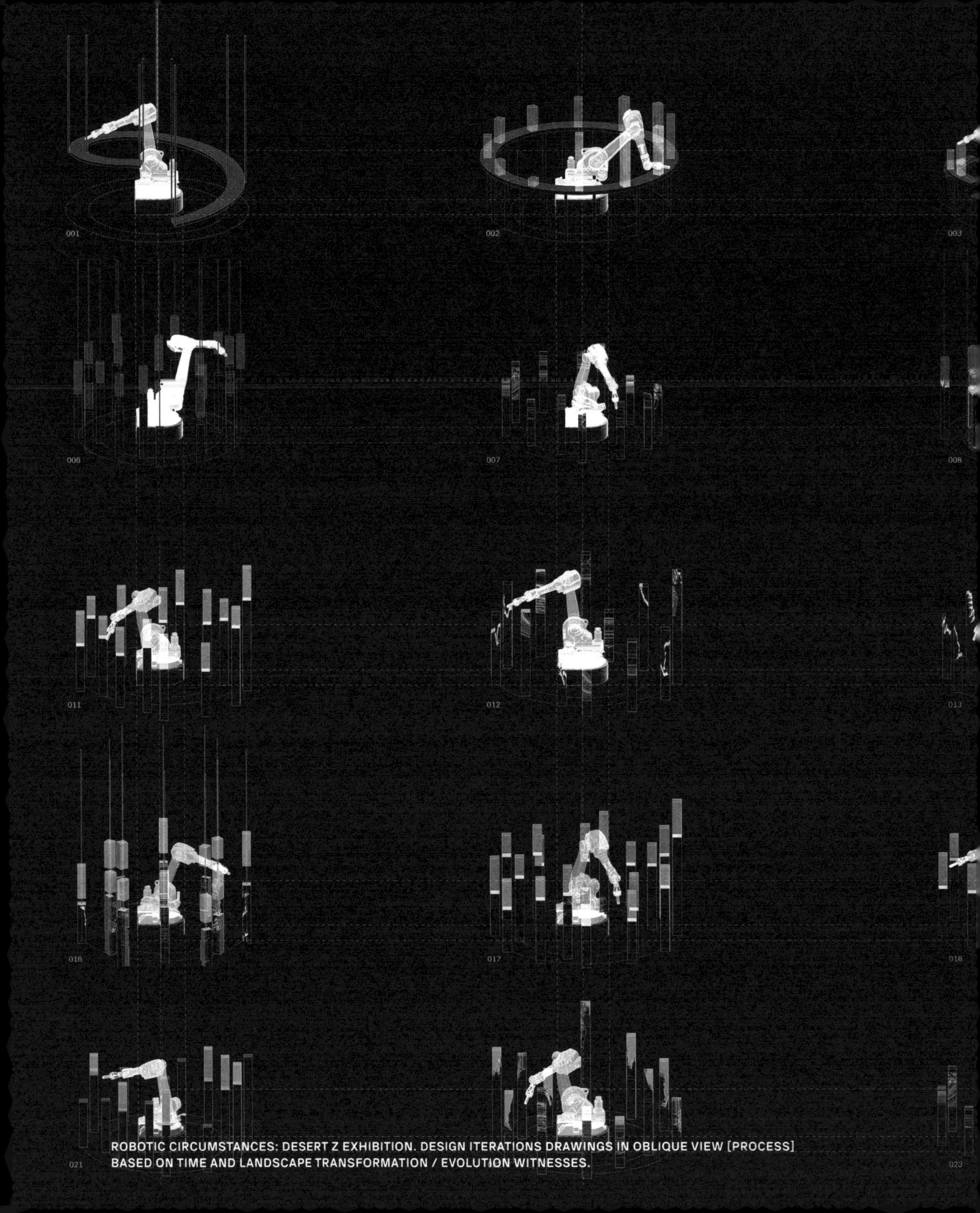

ROBOTIC CIRCUMSTANCES: DESERT Z EXHIBITION. DESIGN ITERATIONS DRAWINGS IN OBLIQUE VIEW [PROCESS]
BASED ON TIME AND LANDSCAPE TRANSFORMATION / EVOLUTION WITNESSES.

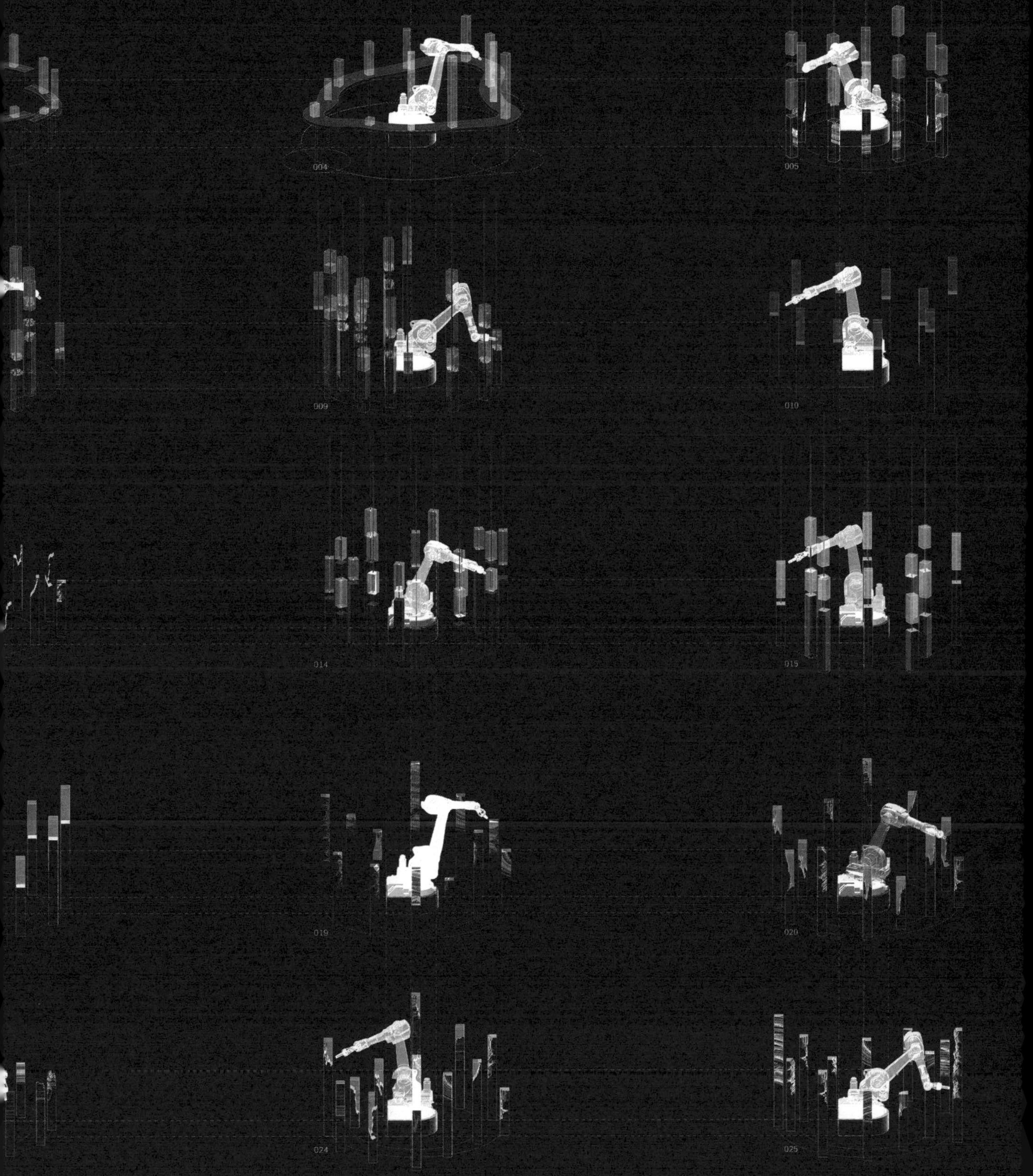
004
005
009
010
014
015
019
020
024
025

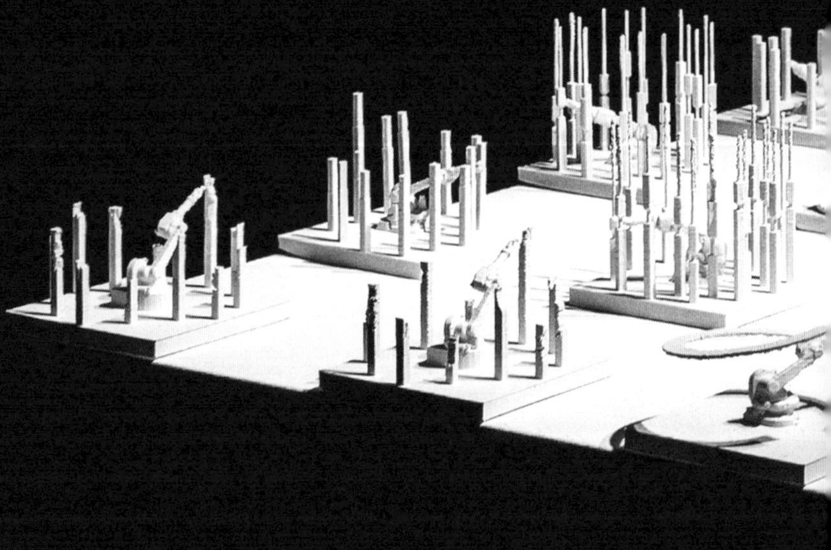

ROBOTIC CIRCUMSTANCES: DESERT Z EXHIBITION. DESIGN ITERATIONS PHYSICAL MODELS [PROCESS]
PRESENTED DURING EXHIBITION. AXONOMETRIC VIEW. FABRICATION: 3DPRINTED + PAINTED. 2023

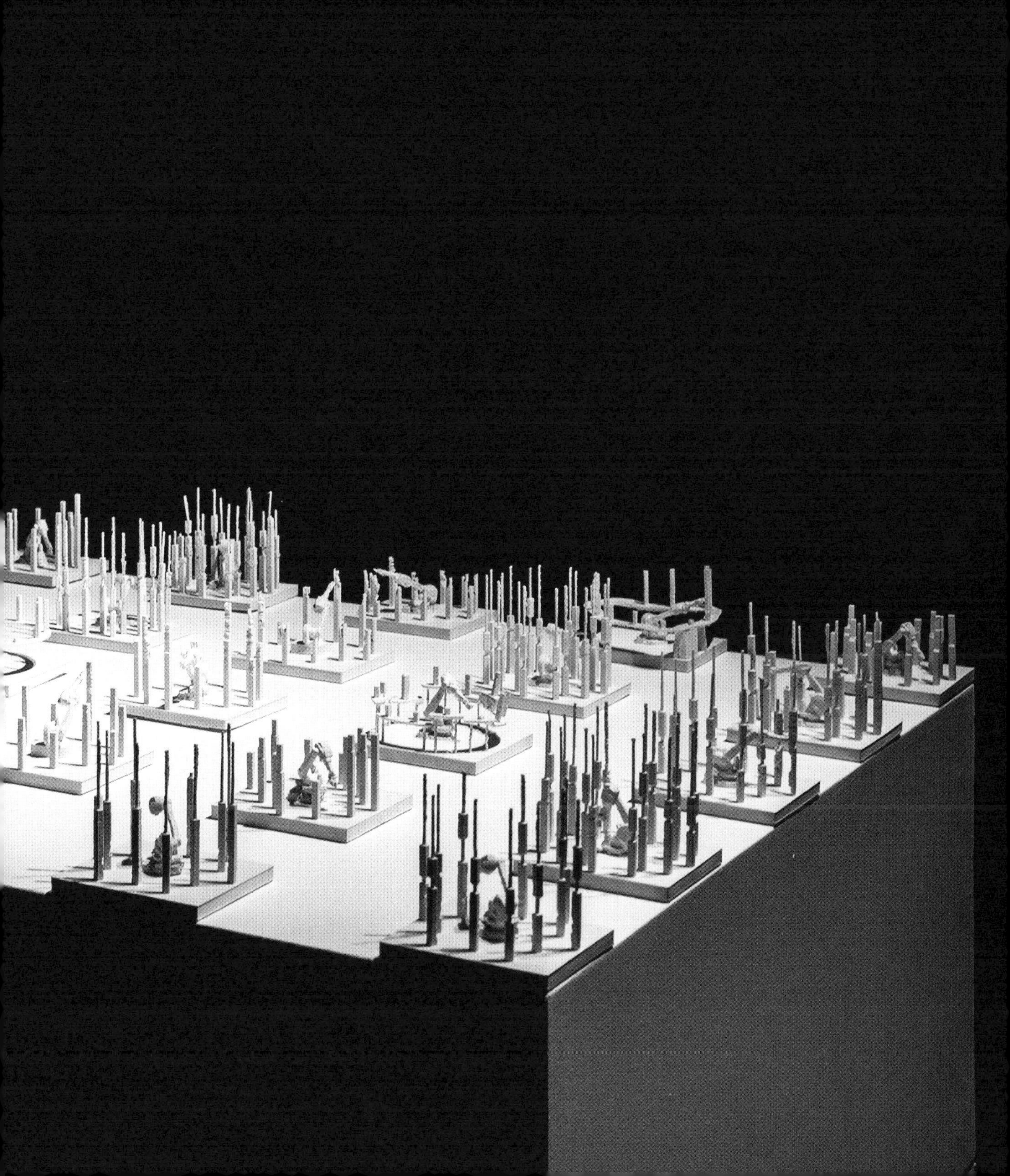

ROBOTIC CIRCUMSTANCES: DESERT Z EXHIBITION. DESIGN ITERATIONS PHYSICAL MODELS [PROCESS] PRESENTED DURING EXHIBITION. CLOSE-UP VIEW. FABRICATION: 3DPRINTED + PAINTED. 2023

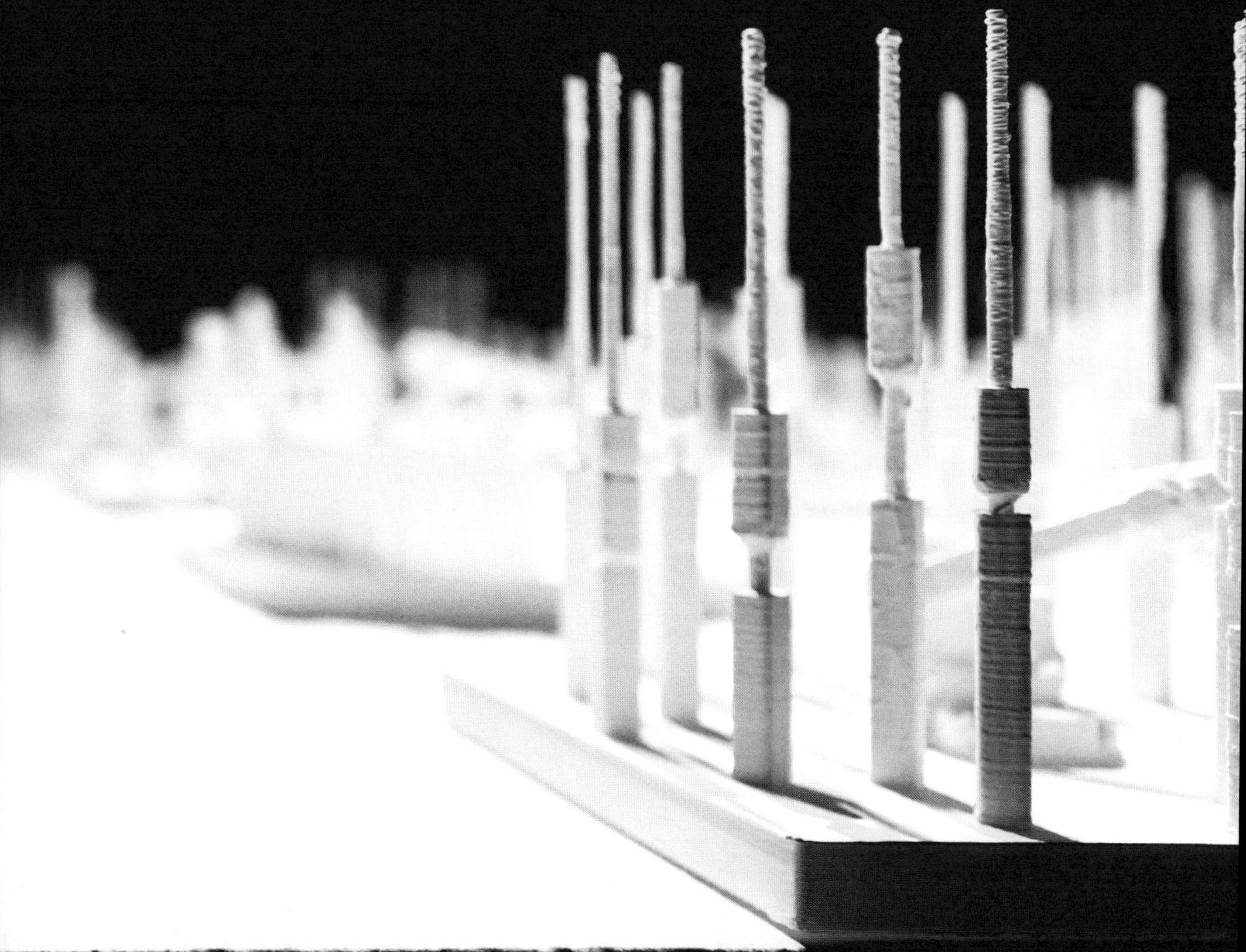

ROBOTIC CIRCUMSTANCES: DESERT Z EXHIBITION. DESIGN ITERATIONS DRAWINGS IN OBLIQUE VIEW.
12 WITNESSES INFORMED BY LANDSCAPE TRANSFORMATION / EVOLUTION.

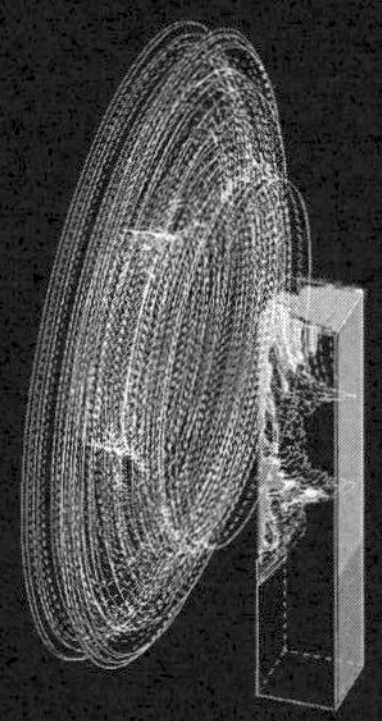

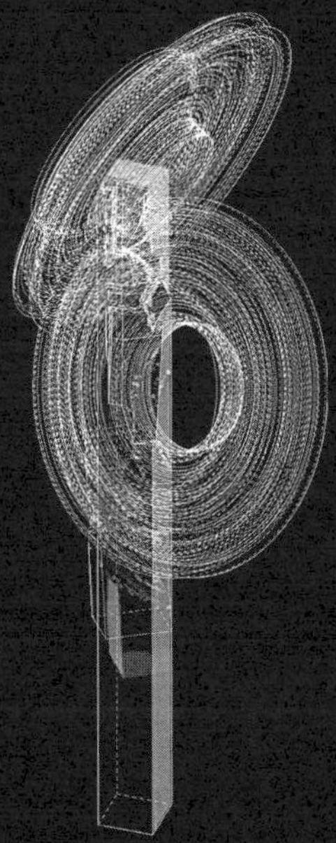

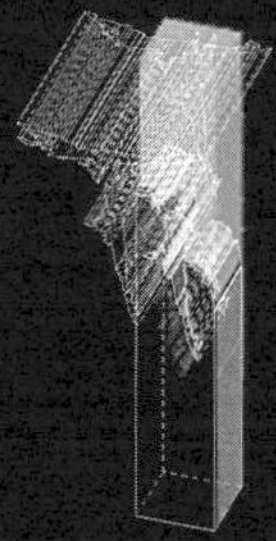

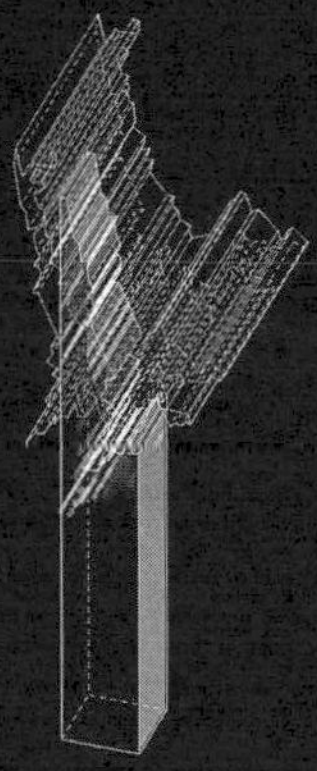

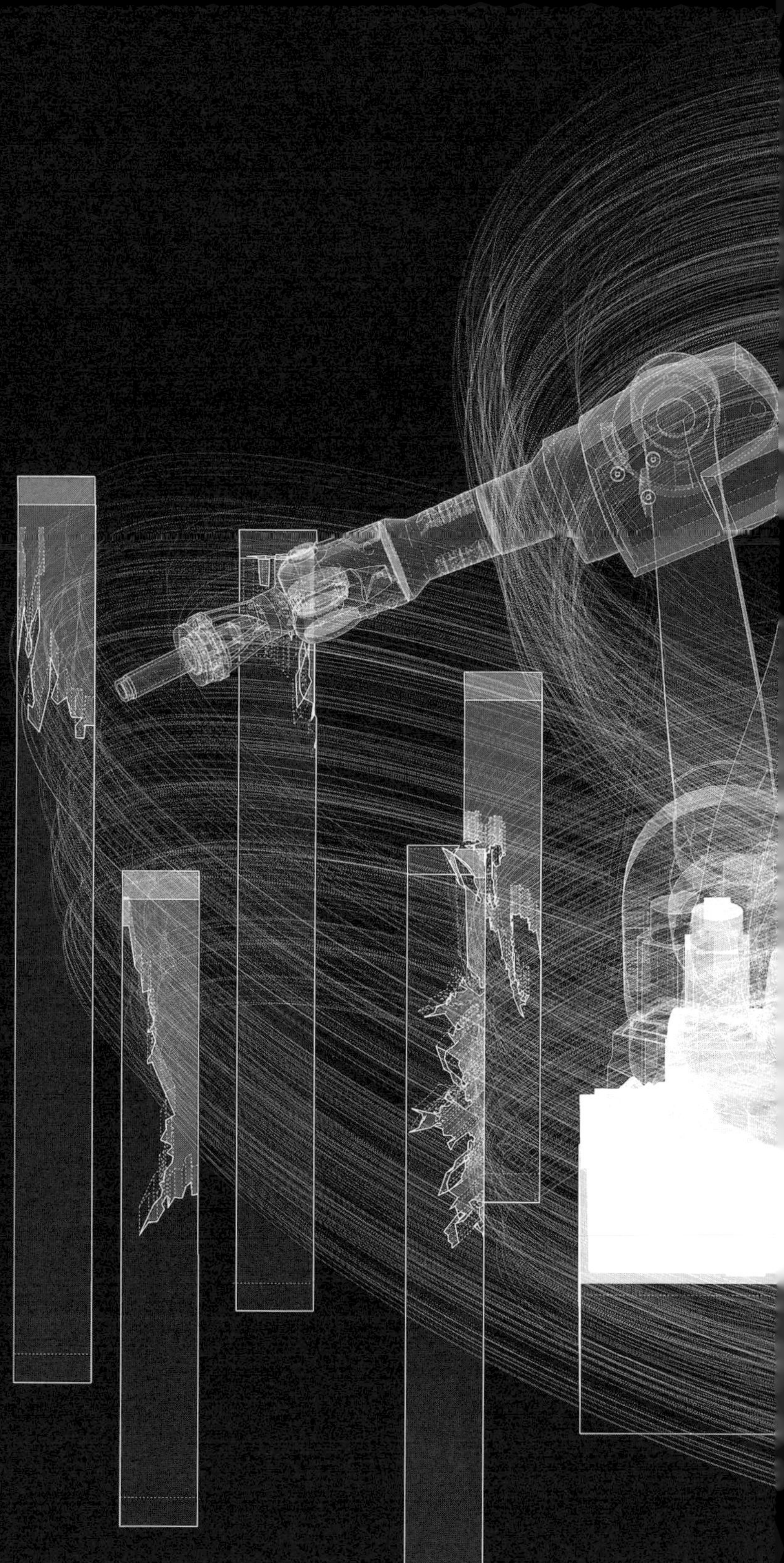

ROBOTIC CIRCUMSTANCES: DESERT Z EXHIBITION. SCENOGRAPHY + CHOREOGRAPHY.
12 WITNESSES TRANSFORMED BY AN ABB 1600 ROBOT + HEATGUN TOOL.

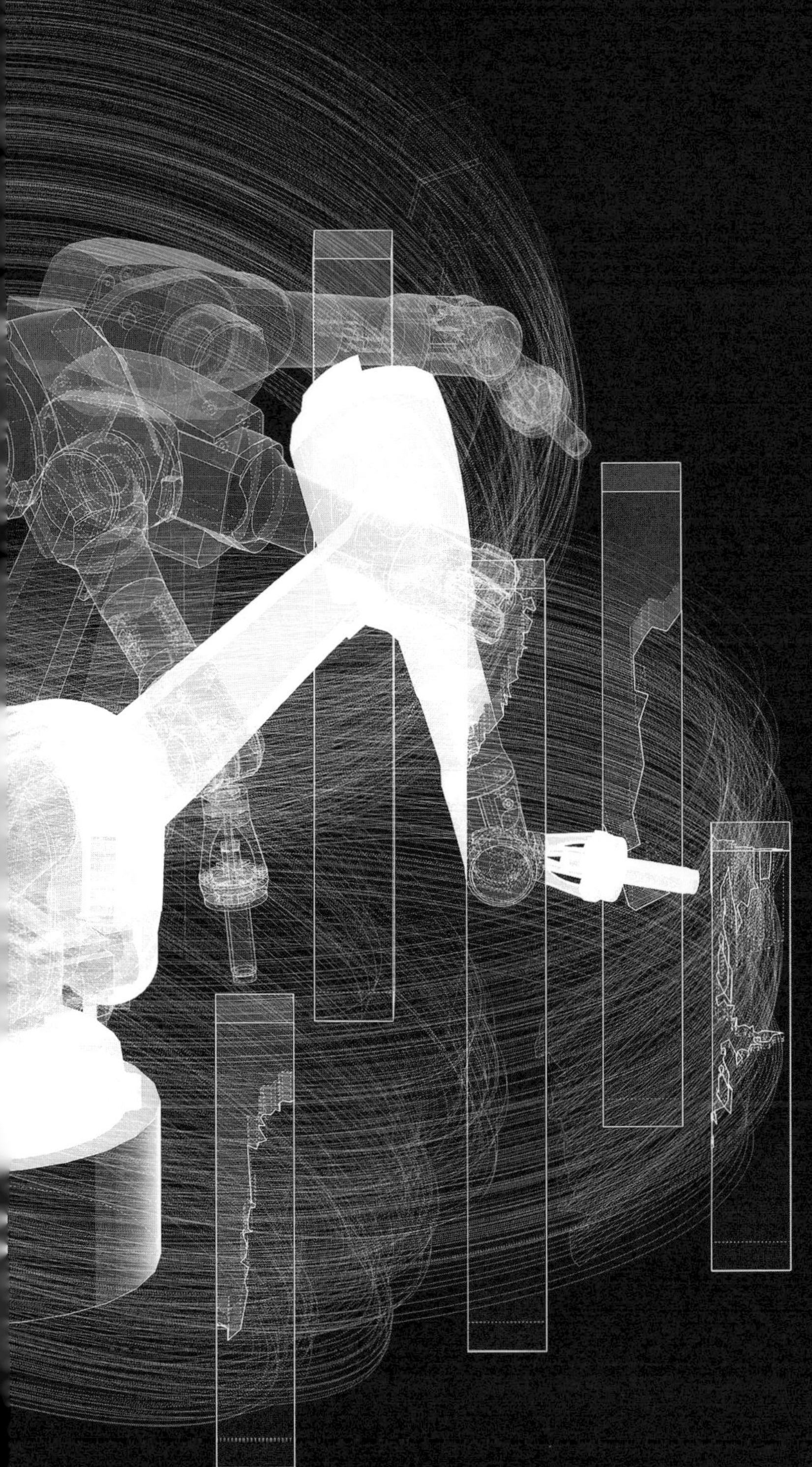

ROBOTIC CIRCUMSTANCES: DESERT Z EXHIBITION. LEFT: DAY 1_9:00AM. RIGHT: DAY 3_6:00PM. 2023

ABB
ABB

ROBOTIC CIRCUMSTANCES: DESERT Z EXHIBITION. LEFT: DAY 2_9:00AM. RIGHT: DAY 2_6:00PM. 2023

ABB
ABB

ROBOTIC CIRCUMSTANCES: DESERT Z EXHIBITION. CLOSE-UP DURING ROBOTIC PERFORMANCE. DAY 1_9AM. 2023

ROBOTIC CIRCUMSTANCES: DESERT Z EXHIBITION. VERTICAL WITNESSES [+Z -Z]. DAY 1_10AM. 2023

ROBOTIC CIRCUMSTANCES: DESERT Z EXHIBITION. CLOSE-UP DURING ROBOTIC PERFORMANCE. DAY 2_10AM. 2023

ROBOTIC CIRCUMSTANCES: DESERT Z EXHIBITION. CLOSE-UP DURING ROBOTIC PERFORMANCE. DAY 2_5PM. 2023

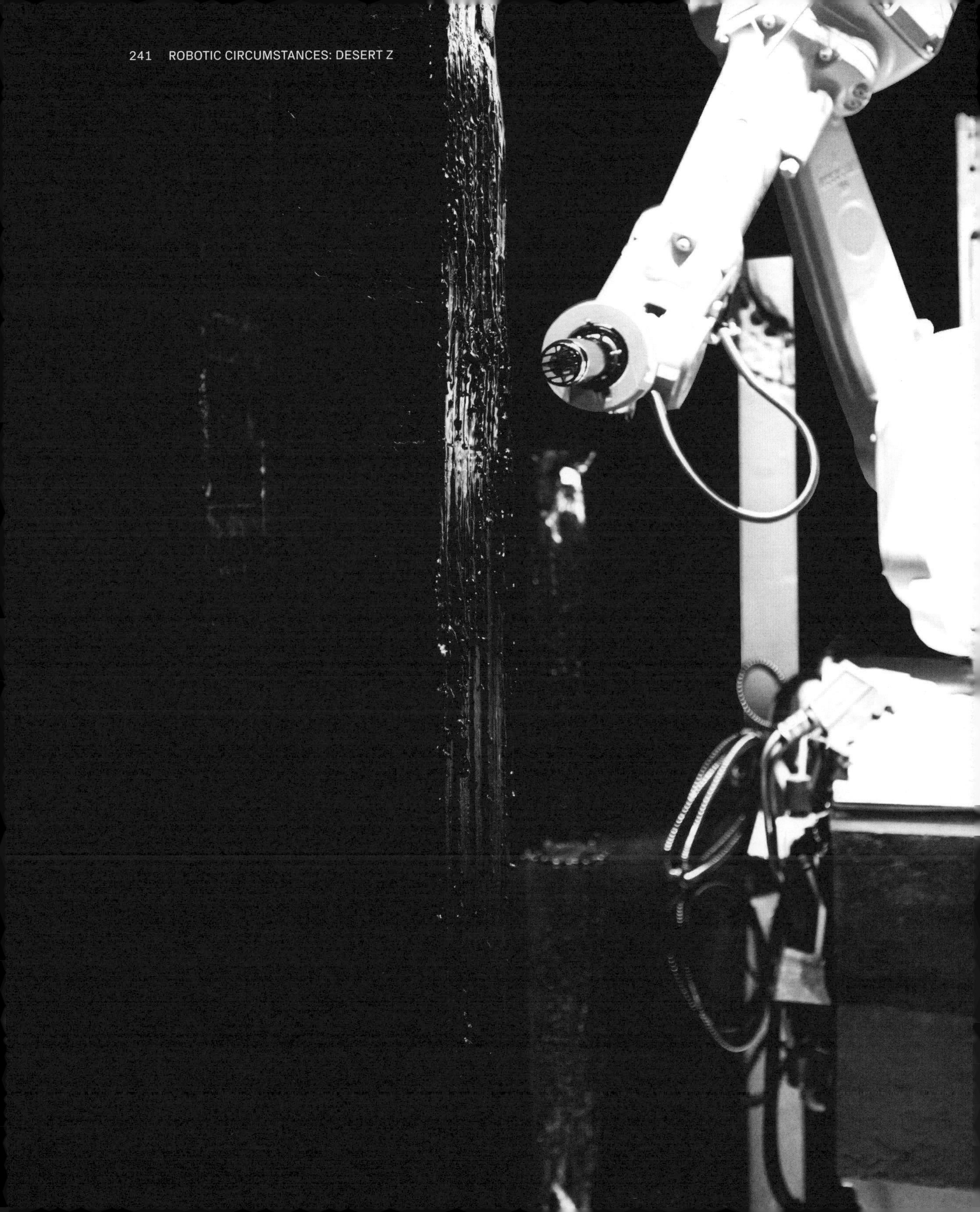

ROBOTIC CIRCUMSTANCES: DESERT Z EXHIBITION. CLOSE-UP DURING ROBOTIC PERFORMANCE [BACKSTAGE]. DAY 2_9AM. 2023

ROBOTIC CIRCUMSTANCES: DESERT Z EXHIBITION. AERIAL VIEW. DAY 3_6PM. 2023

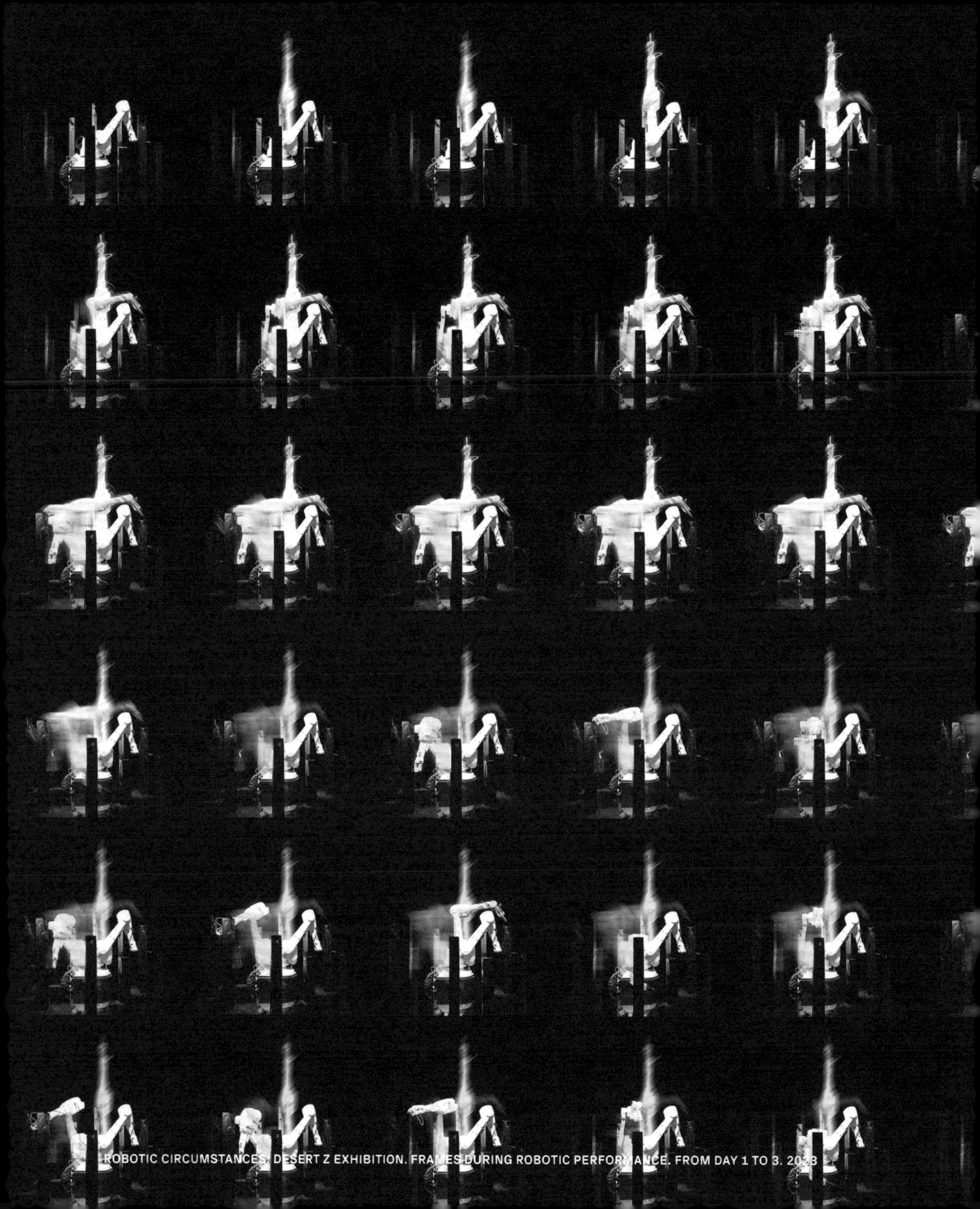

ROBOTIC CIRCUMSTANCES. DESERT Z EXHIBITION. FRAMES DURING ROBOTIC PERFORMANCE. FROM DAY 1 TO 3. 2023

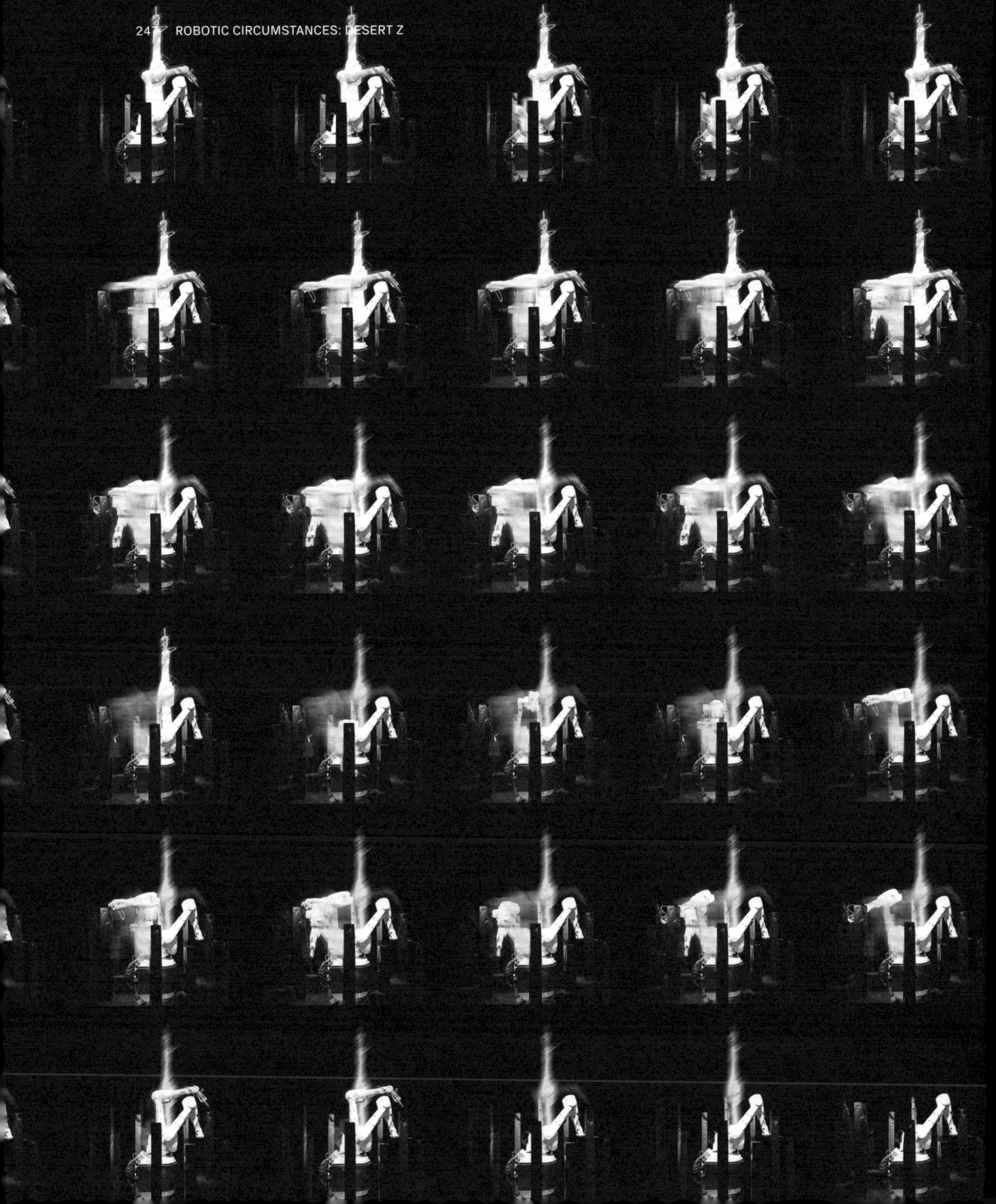

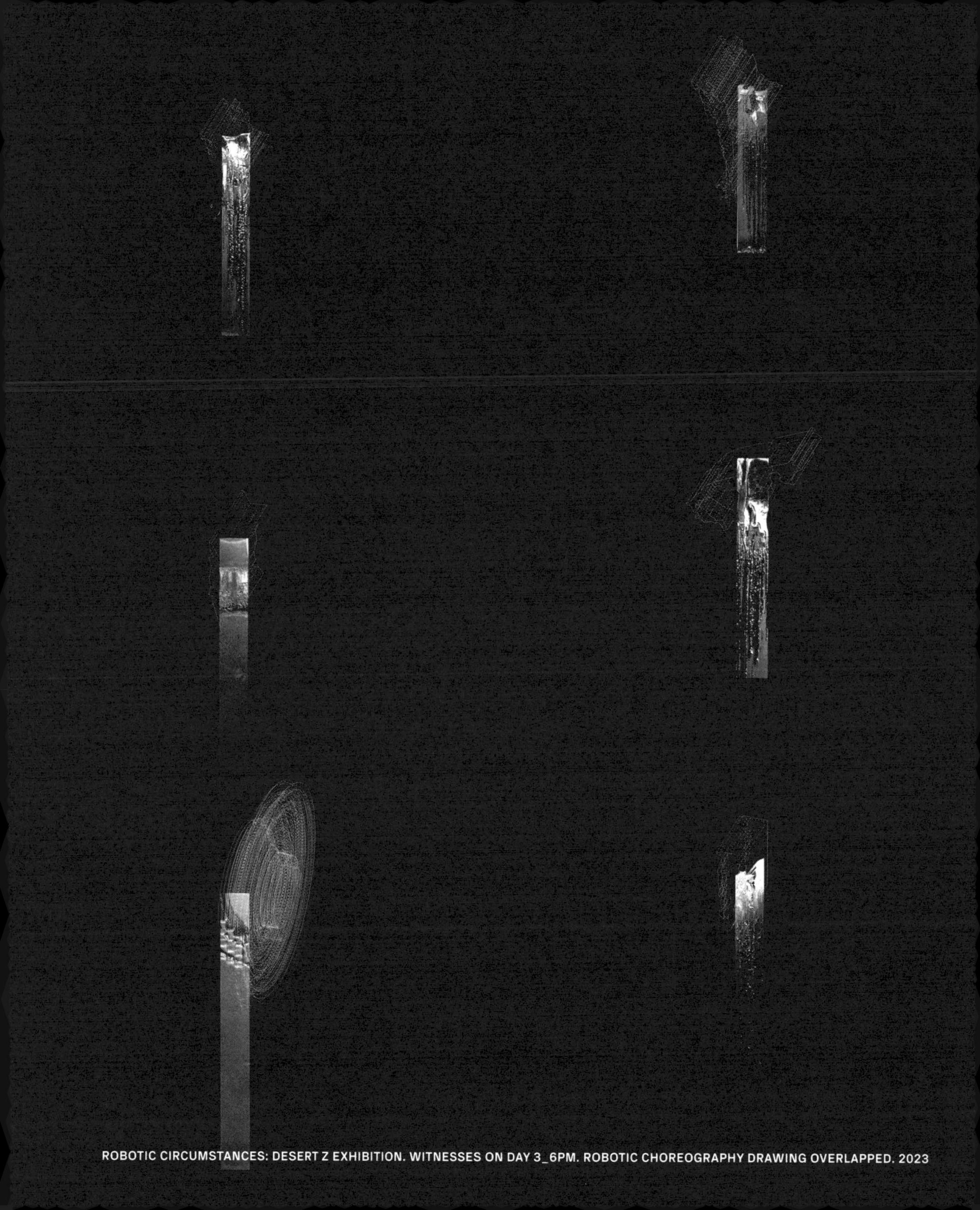

ROBOTIC CIRCUMSTANCES: DESERT Z EXHIBITION. WITNESSES ON DAY 3_6PM. ROBOTIC CHOREOGRAPHY DRAWING OVERLAPPED. 2023

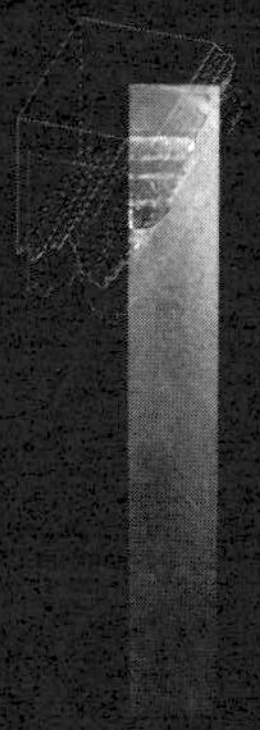

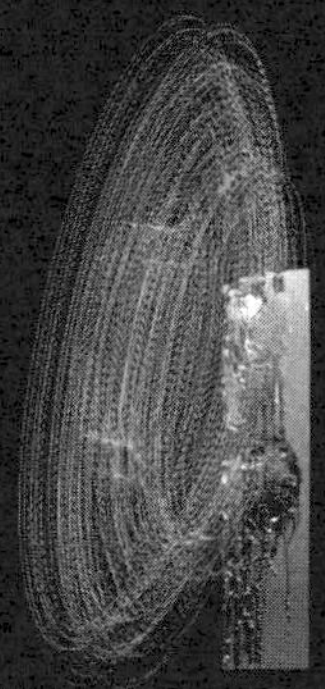

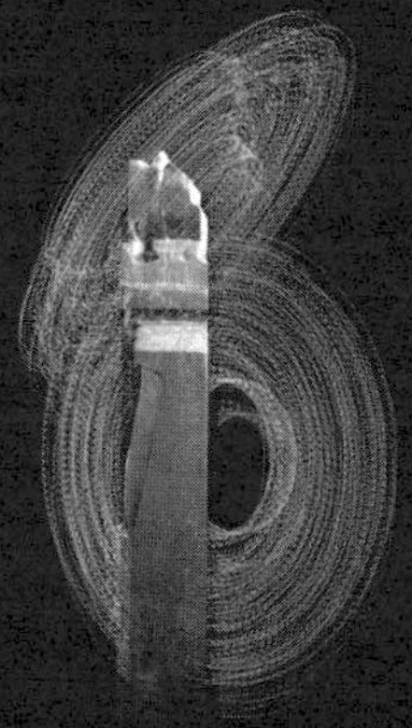

ROBOTIC CIRCUMSTANCES: DESERT Z EXHIBITION. MATERIAL TRANSFORMATION / EVOLUTION DURING EXHIBITION. 2023

ROBOTIC CIRCUMSTANCES: DESERT Z EXHIBITION. MATERIAL TRANSFORMATION / EVOLUTION CLOSE-UP ON DAY 3_6PM. 2023

ROBOTIC CIRCUMSTANCES: DESERT Z EXHIBITION. SCENE DURING ROBOTIC PERFORMANCE. 2023

ROBOTIC CIRCUMSTANCES: DESERT Z EXHIBITION. SCENE DURING ROBOTIC PERFORMANCE. 2023

ROBOTIC CIRCUMSTANCES: DESERT Z EXHIBITION. MATERIAL CLOSE-UP DURING EXHIBITION. 2023

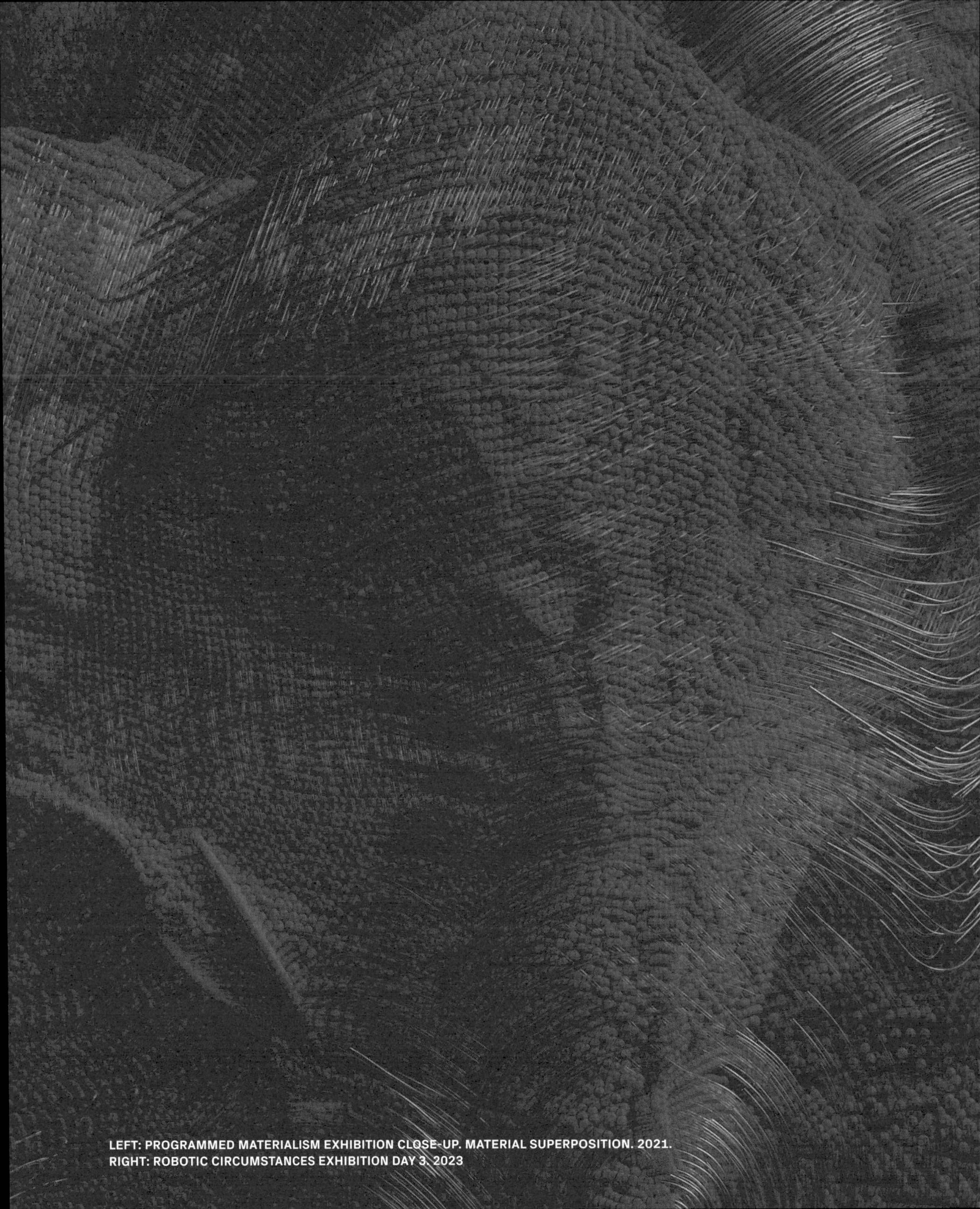

LEFT: PROGRAMMED MATERIALISM EXHIBITION CLOSE-UP. MATERIAL SUPERPOSITION. 2021.
RIGHT: ROBOTIC CIRCUMSTANCES EXHIBITION DAY 3. 2023

LEFT: PROGRAMMED MATERIALISM EXHIBITION CLOSE-UP. NEW MATERIALISM. 2021.
RIGHT: ROBOTIC [PHYSICAL] TRANSLATIONS. TOOL: MOTOTOOL. 2022

LEFT: PROGRAMMED MATERIALISM EXHIBITION CLOSE-UP. UNLOCKING THE MATERIALITY. 2021.
RIGHT: ROBOTIC CIRCUMSTANCES EXHIBITION DAY 1. 2023

ACKNOWLEDGEMENTS

This book has been a labor of passion and dedication, and I am profoundly grateful to all those who have contributed to its realization. First, I am indebted to the contributors and assistants over the years.

To Claudio Rossi my partner in this adventure. Your solid support and constant intellectual, and knowledgeable feedback are part of many of the debates that cross the continents in the structure of this book. Also, to complement my gaze of Latin-American context, technology, architecture, teaching, and research, in Worm-Eye View, but particularly for sharing with me the answer to the question of: Why Black?

To Gabriel Esquivel and Peter Testa my admiration and the satisfaction of having been able to connect bridges between your contributions and the results that today are shown in this book.

To the Vice-rectory of Research and Creation of the Universidad de los Andes, with its research-creation grant and the Faculty of Architecture and Design (ArqDis) working platform around the robot including the support of software and other machines (hardware), that helped me to structure and execute the multiple products that were derived from the *Programmed Materialism* research-creation project including this book; but also, the libraries and digital platforms that provided invaluable resources and support throughout the research process. It's a pleasure to be part of their commitment to preserving knowledge and fostering intellectual inquiry. It is important to recognize the work of the people behind the operation of the ArqDis for their support in the various phases of this process, from the Dean Hernando Barragán, and the Chair Claudia Mejía, to the various instances in the navigation of a project of this type, both logistical and resources, consulting, licensing, and registration. Thanks for supporting of this pioneer research-creation project in our Department, Faculty, University and Region.

To the team members of the research-creation seedbed Video Games & Robots formulated by Claudio Rossi and myself few years ago, with the purpose of developing workflows from technology as strategies for knowledge production from early stages of the architect's formation, and as a space connected to Robot-Lab, the academic engine that tested all the innovative pedagogies presented here. As assistants in this research-creation team, I want to express special thanks to Valeria Ayala, Camila Dussan and Daniela Duarte (undergrad students from Facultad de Arquitectura y Diseño, Universidad de Los Andes) that helped me gathering together and selecting part of the info for this book.

Other team members and equally important like Valentina Trujillo, Valentina Aguilar, Daniela Duarte, Felipe Zorro, Laura García, and Sofía Prado (undergrad students from Facultad de Arquitectura y Diseño, Universidad de Los Andes) that were collaborators in the Desert-Z exhibition. Also, team members that were part of this adventure since the beginning: Lina Triviño, Daniel Barrera, and Juan David Frank. Also, to the Robot_Lab course Teaching Assistants: Valentina Trujillo, Juan Felipe Castiblanco, Juan David Urquina, Gabriel Niño, Mariana Aellos, Emma Bonilla, and Andrea Rodríguez, all of them now alumni of Universidad de Los Andes.

Programmed Materialism also helped to the create and grow and ecosystem that involves the technicians of the Universidad de Los Andes workshop. To all of them many thanks for letting me lead a plural, creative, and innovative group that pursued ideas for the future and disruptive positions about our Latin American context.

To all the Robot-Lab students during its operation (2021-2023) for taking the risk with Claudio and me, but also, for believe in the idea of an alternative position to approach the wonderful reality (magical realism) of the Latin American region, especially from/with the use of technology, in particular, the robotic arm.

I would like to also thank Casey Rehm, Coordinator of the Master of Architectural Technologies at Southern California Institute of Architecture (SCI_Arc), for welcoming and letting us work with the "Giant Robot", but also for accompanying this adventure in the desert with significant lessons to the debate of this research.

Finally, I extend my deepest thanks to the team at ACTAR Publishers, in the hand of Ricardo Devesa, whose dedication, professionalism, and expertise have made this publication possible. Their innovative spirit and commitment to excellence (and the patience) have been instrumental in bringing this project to be in my (and your) hands. Thanks for the support and *materialization* of this 'paper-binded hyper-artifact'.

And, as always, to my family for their support and for constantly *translating* to them everything that I do. They are simply an example of hard work, dedication, coherence, and consistency.

This book would not have been possible without the contributions of each, and every individual mentioned above. While space constraints prevent me from acknowledging everyone by name, please know that your support is deeply appreciated and will always be remembered.

Daniela Atencio

AUTHOR & CONTRIBUTORS

Daniela Atencio. Author

Architect (Summa Cum Laude). She completed her postgraduate degree (Master of Design Research in Emerging Systems, Technologies, and Media) at SCI_Arc in the United States (graduated with Distinction). She is an associate professor and researcher at the University of Los Andes (Bogotá, Colombia), where she coordinates courses in computational thinking, robotics, and final-year projects with a technological emphasis. She developed research projects Programmed Materialism ([ranslations between images, objects, robots, and pedagogies], Hyper_Artifacts, Hyper_Anomalies, and Bio_lenta Architecture. Member of the Women in Robotics Association, where she coordinates the South American chapter. She is co-founder of the Videogames & Robots research group at the University of Los Andes.

Her work covers topics of advanced technologies in architecture and design, representation, computational design, robotics in architecture [computer controlled], published and/or exhibited by platforms, associations, and publishers, such as the 17th Venice Architecture Biennale Italian Pavilion Sezione del Padiglione Italia, the International Association of Robots in Architecture [ROB | ARCH], Association for Computer Aided Design in Architecture [ACADIA]: REALIGNMENTS Toward Critical Computation, DigitalFutures [Robotics and Digital Manufacturing in Latin America and Robotics & 3DPrinting], Sociedad Iberoamericana de Gráfica Digital [Sigradi]: Appropriation of Digital Fabrication from Latin America XXVI, XII Ibero-American Biennial of Architecture and Urbanism, Routledge, AD Journal, Robotic circumstances: DESERT Z [Real-time robotic exhibition for Pavilion 2023], among others. In 2021, along with Prof. Claudio Rossi, started the first robotics course in undergraduate architecture in Latin-America called Robot_Lab.

She has been invited as a jury and/or lecturer at SCI_Arc, UCLA, University of Melbourne, Pontificia Universidad Católica de Chile, Kent State University, Texas A&M, Universidad de Monterrey, Universidad del Rosario [Argentina], The University of Manchester, Manchester Metropolitan University, among others.

Claudio Rossi. Editor & Contributor

Doctor in Architecture / Master in Urban Design / Architect / with more than 20 years of experience in architectural and urban projects that have been awarded, such as the Urban Project for the Central Coast of Venezuela after the Natural Disasters of 1999 [National Architecture Award and Honorable Mention for the Best Urban Project, Caracas Biennial]. Awarded in different competitions and invited to be: lecturer and visiting professor in Architecture, Urban Design, and Landscape Architecture at Politecnico di Torino, IT [2018 – 2019]. Final Critic Reviewer, PennDesign, US [Dec-2017] / Visiting Researcher, Politecnico di Torino, IT [Jun-2017] / Final Critic Reviewer: Urban Stitching, MIT, US [2016] / Workshop Professor: Limapolis, Pontificia Universidad Católica del Perú, PE [2012-2017] / Workshop Professor: Urban Thickness, Universidad Nacional de Córdoba, AR [Mar-2012] / Professor: International Architecture Workshop of Cartagena, COL [2010-2014, 2018] / Speaker, International Congress: Cities in Transformation at Politecnico di Milano [2012] / Coordinator of the Master of Architecture at Universidad de Los Andes [2013-2017] / among others.

Associate Professor, Faculty of Architecture and Design at Universidad de Los Andes, COL; Interested in the contemporary responsibilities of the architectural project and its position in the Latin American landscape. Working with technology to expand the arguments and narratives of the project and its agency trough representations.

His research topics and projects are: architectural anomalies, bio-lent architecture and projectual inflexions in Latin America, collisions between landscape and infrastructure, resilient territories and the representation of the Latin American landscape in video games.

Peter Testa. Contributor

Peter Testa is a founding partner and design principal of TestaWeiser, located in Los Angeles. He is also a senior design faculty member at SCI-Arc. At TestaWeiser, he leads strategy and design for some of the world's most innovative companies and government agencies. His work has been featured at renowned institutions such as the Los Angeles Museum of Contemporary Art (MOCA), The Centre Pompidou, The Cooper Hewitt Smithsonian Design Museum, and the National Museum of Modern Art in Tokyo. Additionally, his work is held in public collections and forms part of the permanent collection of the Canadian Centre for Architecture.

In 2016, his firm's Carbon Tower was recognized as one of twenty canonical projects of the digital turn in the exhibition 'Archeology of the Digital' at the Canadian Centre for Architecture and Yale School of Architecture. His projects and writings have been published in international art, architecture, design, and scientific journals, as well as major newspapers. His most recent book, "Robot House," was published by Thames & Hudson in 2017.

Peter Testa has held professorships at Columbia GSAPP and the Massachusetts Institute of Technology and has also served as a visiting professor at several schools including Harvard GSD and the University of California, Berkeley. In 2020, he was awarded the DAAD German Academic Exchange Award and served as a visiting professor at the Städelschule Art Academy in Frankfurt. He holds a B.Arch and an S.M.Arch.S (History/Theory/Criticism) from the Massachusetts Institute of Technology, and is a Registered Architect in California. Prior to founding TestaWeiser he was principal in charge with the office of Álvaro Siza. Throughout his career, he has received numerous awards, including the MIT Innovation Award, the Architecture Award of the Municipal Art Society of New York, and the Design Arts Award of the National Endowment for the Arts.

Gabriel Esquivel. Contributor

Gabriel Esquivel was born and educated as an architect in Mexico City with a degree from the National University and received his master's degree in Architecture from The Ohio State University. Gabriel Esquivel teaches architecture at the College of Architecture at Texas A&M University. He previously taught Architecture and Design at the Ohio State University. He is an invited professor at The Tecnológico of Monterrey in Mexico City since 2010.

After joining the architecture faculty at Texas A&M University, he investigated the benefits and vehicles of a heterogeneous model that integrates both design technology and architecture's proprietary devices. Specifically, Gabriel Esquivel examines digital geometry the emergence of new material logic, and new design methodologies based on Artificial Intelligence. He also examined the integration of digital techniques and analog conventions to exchange architectural information. Consequently, Gabriel Esquivel has taught several studios and workshops in the US, Mexico, and South America that dealt with these issues providing the background for architecture education to be at the top of cutting-edge technology.

In 2011 he created the T4T Lab which is an internationally known Vertical Studio/Lab that invites young practitioners from all over the world to teach together. The lab focuses on the connection between Technology and contemporary theory. He has also developed an Advanced Research fabrication Lab that deals with the concept of Robotic-assisted fabrication and Artificial Intelligence. These projects have been discussed in papers from SMI and Acadia from the parametric point of view as well as theory-based publications like Thresholds from MIT. He is the director of The T4T LAB at Texas A&M (Technology and Theory), the mission of the lab is to invite young practitioners from all over the world to develop their ideas within a lab/studio format. He created the AI and fabrication lab in 2020 and worked on several projects like the Serlio Code which has been featured in different publications like AD and exhibited in different venues like the Wedge Gallery at the Woodbury School of Architecture.

BIBLIOGRAPHY

Prologue

Atencio, D., Rossi, C. (2024) "*Programmed Ruination. Hyper_Artifacts*". Cit. In: Edited by Esquivel, E. (2024) "*Design Technology and Digital Production. An Architecture Anthology*". London: Routledge 2024.

Baudrillard, J. (2020) "*The System of Objects*". Translated by Benedict, J. London: Verso

Bennett, J. (2010). "*Vibrant Matter. A Political Ecology of Things*". Durham: Duke University Press.

Coeckelbergh, M. (2010) "*Robot Rights? Towards A Social-Relational Justification of Moral Consideration*". Ethics Inf Technol 12.

De Pagter, J. (2021) "*Frontiers. Speculating About Robot Moral Standing: On the Constitution of Social Robots as Objects of Governance*". December 2, 2021. www.frontiersin.org - Institute for Management Science, TU Wien, Vienna, Austria.

Herrera, P., Juárez, B. (2012) "*Perspectives in Latin America's Fab Labs*". Universidad Peruana de Ciencias Aplicadas, Perú. Espacios Digitales. Fab Lab Lima.

Hutchins, E. (1996) "*Cognition in the Wild*". Massachusset: The MIT Press.

Latour, B. (2005) "*Reassembling the Social: An Introduction to Actor-Network-Theory*". New York: Oxford University Press.

Merrihew Adams, R. (2022) "*What Is, and What Is In Itself. A Systematic Ontology*". New York: Oxford University Press.

Porter, E. "*Don't Fight The Robots. Tax Them*", New York Times; February 23, 2019, https://www.nytimes.com/2019/02/23/sunday-review/tax-artificial-intelligence.html

Robotic Translations: Design Processes

Alberti, L.B. (1485) "*De re ædificatoria: Vols. VI–XIX*". Translated by Rykwert, J., Tavernor, R. & Leach, N. (1988), Cambridge: MIT Press.

Atencio, D. (2021) "*Programmed Materialism. Fictional Translations*". 17th Venice Architecture Biennale Italian Pavilion Sezione del Padiglione Italia—CITYX, Venice.

Atencio, D., & Rossi, C. (2023) "*Robot-Lab: Hyper-Artifacts*". Catalog exhibition: Homo Faber 3.0 Appropriations of Digital Fabrication from Latin America, Lima: Editorial UPC.

Benedetti, M. "*El Sur también Existe*". Poems.

Bergson, H. (1896) "*Of the Selection of Images for Conscious Presentation. What Our Body Means and Does*". Translated by N.M. Paul, N.M., & Palmer, W.S. (1911) "*Matter and Memory*". London: George Allen and Unwin.

Bilbao, T., Greenberg, & N., Ghosh, A. (2020) "*Two Sides of the Border. Reimagining the Region*". Zurich: Lars Müller Publishers.

Braudel, F. (1982) "*Civilization and Capitalism, 15th-18th Century, Volume I: The Structures of Everyday Life*". Berkeley, University of California Press

Braudel, F. (1958) "*History and the Social Sciences: The Longue Durée*". Review Journal, Research Foundation of State University of New York, 32 (2) 2009.

Cardozo Galué, G. (2013) "*Significaciones y Alcances del Concepto Región Histórica*". Cambios & Permanencias (4).

Cardozo Galué, G. (1998) "*La Región en el Proceso Histórico Venezolano. Propuesta de Periodización para la Región de Maracaibo*".

Caravelle. Cahiers du Monde Hispanique et Luso-Brésilien, 70

Carpo, M. (2014) "*Breaking the Curve: Big Data and Design*", Artforum.

Carpo, M. (2017) "*The Second Digital Turn*", Cambridge: The MIT Press.

Carrasquero, A., Finol, J., & García, N. (2009) "*Símbolos, Espacio y Cuerpo en la Yonna Wayuu*". Revista de Ciencias Sociales, 15(4).

Ciriani, E. (2014) "*Todavía la Arquitectura*". Lima: Editorial Arcadia.

Clear, N. (2014) "*Convergence: Architecture as Integrated Spatial Design*". Cit. In: Spiller, N., & Clear, N., "*Educating Architects: How Tomorrow's Practitioners Will Learn Today*". London: Thames & Hudson.

Clément, G. (1991) "*Le Jardin en Mouvement*". Paris: Pandora.

Colina, S. (2015) "*Fundamentals of Translations*". Cambridge: Cambridge University Press.

De Vaan, M. (2008) "*Etymological Dictionary of Latin and the other Italic Languages*", Leiden: Boston.

Deleuze, G. (1983) "*Cinéma 1. L'image-mouvement*". Paris: Les Éditions de Minuit.

Delderfield, R. F. (1945) "*Worm-Eye View*". Theater play.

Dictionnaire de L'Académie Française, 8e éd., [online ver.]. https://www.dictionnaire-academie.fr/article/A8S1966

Durand, J.N.L. (1802-5) "*Précis des Leçons D'architecture Données à l'École Royale Polytechnique*". Paris: The Author.

Eisenman, P. (1998). "*Diagram: An Original Scene of Writing*". ANY: Architecture New York, 23.

Elden, S. (2009) "*Terror and Territory*". Minneapolis: Minnesota University Press.

Elden, S. (2013) "*The Birth of Territory*". Chicago: Chicago University Press.

Ferrater Mora, J. (1994) "*Diccionario de Filosofía*", Madrid: Editorial Ariel.

Galloway, A. (2014) "*Laurelle Against the Digital*". PostHumanities 31, Minneapolis: University of Minnesota Press.

García Márquez, G. (2000). Interview for the newspaper La Reforma de México.

Goshn, R., Jazairy, EH. (2020) "*Geostories: Another Architecture for the Environment*". Barcelona: ACTAR.

Heidegger, M. (1956) "*Construir, Habitar, Pensar*". Trad. (2015) Jesús Adrián Escudero. Barcelona: Oficina de Arte y Ediciones.

Henni S. (2022) "*Deserts are Not Empty*". New York: Columbia University Press.

Latour, B. (2007) "*Could We Have our Materialism Back Please?*", In Thick Things edited by Ken Alder.

Laurelle, F. (1996) "*Principles of Non-Philosophy*", London: Bloomsbury Academic.

Lawrence, R. (1963) "*Machine Perception Of Three-Dimensional Solids*" Thesis (Ph. D.) Massachusetts Institute of Technology, Dept. of Electrical Engineering.

Le Corbusier (1929) "*Tracés Régulateurs*" cit. in. Quetglas, J. (2009) "Les Heures Claires. Proyecto y Arquitectura en la Ville Savoye de Le Corbusier y Pierre Jeannerete". Barcelona: Asimétricas.

Lewis, P., Tsurumaki, M., & Lewis, M. (2016) "*Manual of a Section*". New York: Princeton Architectural Press.

Limerick, P. (1985) "*Desert Passages. Encounters with the American Deserts*". Albuquerque: University of New Mexico Press.

Loos, A. ([1910] 2019) "*Ornament and Crime: Thoughts on Design and Materials*", New York: Penguin.

Lotman, J. (2005) "*On the Semiosphere*". Translated by Clark, W. Sign in Systems Studies, 33.1.

Merleau-Pounty, M. (1968) "*The Visible and the Invisible*". Evanston: Northwestern University Press.

Mostafavi, M., & Leatherbarrow, D. (1993) "*On Weathering. The Life of Buildings in Time*". Cambridge: MIT Press.

Muholi, Z., Mussai, R., & Azzua, U. (2018) "*Zanele Muholi: Somnyama Ngonyama, Hail the Dark Lioness*". New York: Aperture.

Nahm, M. (1956) "*The Artist as Creator: An Essay of Human Freedom*", Baltimore: Johns Hopkins Press.

Negarestani, R. (2014) "*The Labor of the Inhuman. Part 1: Human*". Brooklyn: E-Flux Journal, 52.

Negarestani, R. (2008) "*Cyclonopedia: Complicity with Anonymous Materials (Anomaly)*". cit. in. Galloway, A. (2014) "*Laurelle Against the Digital*". PostHumanities 31, Minneapolis: University of Minnesota Press.

Negroponte, N. (1995) "*Being Digital*". New York: Knopf Doubleday Publishing Group.

Ortega y Gasset, J. ([1914] 1995) "*Meditaciones del Quijote*". (3.ª ed.) Madrid: Cátedra.

Oxford English Dictionary, s.v. "*'black-on-white'* in *black* (adj. & n.), sense P.11," February 2023.

Paolini, R., & Gutiérrez, R. (1994) "*El Caribe Fortificado*". Bogotá: Ediciones Uniandes.

Payne, J. (2013) "*Variations on the Disco Ball, or, The Ambivalent Object*", Project 2 (Summer).

Pérez-Gómez, A. (2012) "*The Historical Context of Contemporary Architectural Representation*", cit. in. Ayres, P. (Ed.). "*Persistent Modelling: Extending the Role of Architectural Representation*" (1st ed.). London: Routledge.

Pozzo, A. (1693) "*Perspectiva Pictorum et Architectorum, Vol. 2*". Roma: Typis Joannis Jacobi Komarek.

Quaroni, L. (1977) "Proyectar un Edificio: Ocho Lecciones de Arquitectura". Madrid: Xarait Ediciones

Rael, R. (2017) "*Borderwall as Architecture: A Manifesto for the U.S.-Mexico Boundary*". Berkeley: University of California Press.

Real Academia Española: Diccionario de la Lengua Española, 23.ª ed., [versión 23.6 en línea]. <https://dle.rae.es>

Reinhardt, A. (1970) "*Black Paintings, 1951-1967*". Edited by Rose, B., & Arnason, H. New York: Marlborough Gallery, Inc.

Rossi C. (2021) "*Límites, Bordes y Fronteras*". In: "*Grandes Lecciones, Pequeñas Edificaciones*". Bogotá: Ediciones Uniandes.

Shelley, M. W. (1818) "*Frankenstein; or, The modern Prometheus*". London: Printed for Lackington, Hughes, Harding, Mavor, & Jones.

Sloterdijk, P. (1998) "*Spheres*". Frankfurt: Suhrkamp Verlag.

Stern L. (1790) "*The Works of Laurence Stern: The Life and Opinions of Tristram Shandy*", Gentleman. London: Proprietors.

Note: There are a series of parallel works to the contents of the book that have been reported in scientific and creative spaces, extending the bibliography to technical references or research repositories. The bibliography reported here only represents the books cited in the various texts.

Testa, P. (2017) "*Robot House: The New Wave in Architecture and Robotics*", London: Thames & Hudson.

Testut, J.L., & Latarjet, A. ([1928] 1982) "*Tratado de Anatomía Humana*". Barcelona: Salvat.

Tsukamoto, Y., Kaijima, M. (2007) "*Graphic Anatomy Atelier Bow Wow*". Tokyo: TOTO Publisher.

Watanabe, J. (2005) "*La Piedra Alada*". Valencia: Editorial Pre-Textos.

Weller, R., Hoch, C., & Huang, C. (2017) "*Atlas for the End of the World*". Philadelphia: University of Pennsylvania. http://atlas-for-the-end-of-the-world.com

Wilczek, F. (2016) "*A Beautiful Question: Finding Nature's Deep Design*". London: Penguin Books.

Wulf, A. (2016) "*La Invención de la Naturaleza. El Nuevo Mundo de Alexander von Humboldt*". Madrid: Taurus.

Yilmaz, A. (2016), "*Color of Absence and Presence: Reconsidering Black in Architecture*". COLOR Research and Application. London: Wiley.

ROBOTIC WORKSESSION. TOOL: MOTOTOOL. RECTANGULAR PARALLELEPIPED BLOCK. PROFS. DANIELA ATENCIO AND CLAUDIO ROSSI. 2023

Gun Digest® Book of the M1 GARAND

COLLECTING AND SHOOTING AMERICA'S GREATEST BATTLE RIFLE

BY
PATRICK SWEENEY

Published by

Gun Digest® Books, an imprint of Caribou Media Group, LLC
Gun Digest Media
5583 W. Waterford Ln., Suite D
Appleton WI 54913
gundigest.com

To order books or other products call 920.471.4522 ext. 104
or visit us online at **gundigeststore.com**

ISBN-13: 978-1-951115-94-4

Edited by Corey Graff
Design by Jong Cadelina
Cover by Gene Coo

Printed in the United States of America

10 9 8 7 6 5 4 3 2 1